MECHANICAL DESIGN USING CADD

Other CAD books

Voisinet: INTRODUCTION TO CAD, Second Edition
Voisinet: MECHANICAL CAD LAB MANUAL
Obermeyer: ARCHITECTURAL CAD LAB MANUAL
Voisinet: AUTOCAD MECHANICAL LAB MANUAL
Obermeyer: AUTOCAD ARCHITECTURAL LAB MANUAL
Cascade: CASCADET STUDENT MANUAL
Voisinet: COMPUTER-AIDED DRAFTING AND DESIGN

MECHANICAL DESIGN USING CADD

Donald D. Voisinet
Professor of Engineering Technology
and Coordinator of Design and Drafting
Niagara County Community College
Sanborn, New York

McGRAW-HILL BOOK COMPANY

New York   Atlanta   Dallas   St. Louis   San Francisco
Auckland   Bogotá   Guatemala   Hamburg   Lisbon
London   Madrid   Mexico   Milan   Montreal   New Delhi
Panama   Paris   San Juan   São Paulo   Singapore
Sydney   Tokyo   Toronto

Sponsoring Editor   Stephen M. Zollo
Editing Supervisor   Mitsy Kovacs
Design and Art Supervisor   Caryl Valerie Spinka
Production Supervisor   Mirabel Flores

Text Designer   Suzanne Bennett and Associates
Cover Designer   Ed Smith, Inc.
Technical Art Studio   Fine Line, Inc.

Library of Congress Cataloging-in-Publication Data

Voisinet, Donald D., date
  Mechanical design using CADD / Donald D. Voisinet.
    p.   cm.
  ISBN 0-07-067567-8
  1. Machinery—Design—Data processing.   I. Title.
TJ233.V64   1989   88-11737
621.8'15'0285—dc 19   CIP

The manuscript for this book was processed electronically.

Mechanical Design Using CADD

1 2 3 4 5 6 7 8 9 0   HESHES   8 9 5 4 3 2 1 0 9 8

ISBN 0-07-067567-8

CONTENTS

PREFACE

Mechanical design drafting has been an integral part of technical curricula for many years. Graduates entering industry would apply design concepts and manually produce working drawings. With the evolution of automated drafting, designers and drafters now find themselves engaged in the development of the assembly and details using a CADD system. Technical schools have responded to this situation by providing advanced application-level courses. Students may now apply mechanical design theory directly on the CADD system rather than having to pick it up in the "school of hard knocks." That's what this text is all about. If you learn it as a student, you'll be far ahead of the game when beginning your professional career.

Prior to using this text, you must have completed an introduction to CAD course. You should also have an understanding of algebra, trigonometry, and fundamental mechanical drafting concepts. And, if you basically know your way around a CADD system, you are ready to begin *Mechanical Design Using CADD*. The text uses the AutoCAD command structure. Because of the similarity between the systems, however, you may use any unit at your disposal. A "line is a line" whether it is created using AutoCAD, VersaCAD, CADkey, CADAM, or whatever. In the end you will have achieved the status of a qualified entry-level mechanical design drafter.

The artwork in this text was CADD-generated by a machine design student, Mark Voisinet. After completing the program, he (as well as hundreds of others) has gone on to succeed in a position as an industrial mechanical design drafter. Learn the theory well and apply it fully to the computer. You will be likely, indeed, to find yourself with a very bright, exciting, and rewarding career.

Donald D. Voisinet

ICON GLOSSARY REVIEW

BEFORE	AFTER	COMMAND	DESCRIPTION
		ABSOLUTE COORDINATE	A coordinate point located a horizontal distance (X) and a vertical distance (Y) from the drawing origin).
		ARC	A partial circle is drawn by selecting (1) start/end/angle, (2) start/end/radius, (3) start/midpoint/end, (4) three points among other options.
		ARC TANGENT	A command that will automatically create an arc tangent to two other arcs or circles.
		AREA	A command that will determine and display the area of a polygon and, as desired, subtract any internal smaller regions.
		ARRAY (MINSERT, MULTIPLE COPIES)	A copy function that allows objects to be repeat-copied by keying in a specified X and Y distance (or angle) and number of copies. The pattern may be either rectangular or circular (polar).
		AXIS (X AND Y)	The X axis is the horizontal axis on the graphics monitor. The Y axis is the vertical axis on the graphics monitor. The axes may be marked with ruler lines.
		BASE POINT/ (INSERTION POINT, ORGIN)	The base point of a block is the origin around which it will be inserted or placed.
		BLOCK	A command that assembles lines, arcs, and text into a single drawing entity. Once grouped, they are treated as a single object.
		BREAK (EDIT, PEDIT)	An EDIT command that will erase a portion of any object breaking it into two.
		CENTERLINE (AUTOMATIC)	A command that will determine the center of a circle or arc and automatically draw the centerlines.

BEFORE	AFTER	COMMAND	DESCRIPTION
		CHAMFER	A vertex is replaced by a beveled edge when the two intersecting lines are selected. It may be used to create a hypotenuse (beveled edge) when its length is not known.
		CHANGE	Properties of an entity such as layers, linetypes, pens, elevations, and so on, may be altered. Any object need not be erased and redrawn but simply changed.
		CIRCLES	A circle is drawn by selecting (1) center/radius points, (2) center/key radius coordinate, (3) two opposite points on the circle, or (4) three points on the circumference.
		CLOSE	Any perimeter or polygon may be "closed" by issuing the appropriate key command.
		CONSTRUCTION LINES	Reference lines to assist multiview drawing construction. It is especially useful when a grid pattern cannot be used. The lines are used only as reference and will not appear on the finished plot.
		COORDINATE	A point on the drawing that is designated by the X and Y distance from another point or the drawing origin.
		COPY	A command which allows entities or blocks to be copied at any location.
		CURSOR	An indicating mark on the monitor (CRT) which is controlled by several types of input devices such as: joystick, puck, stylus, etc.
		DIMENSION ANGULAR	The automatic generation of a dimension indicating the angle between two nonparallel lines with the dimension line illustrated as an arc.
		DIMENSION-ASSOCIATIVE (STRETCH)	Automatic revision of a dimension and text when an object is stretched, scaled, or rotated. The new dimension is created in its entirety according to the new size and orientation.

BEFORE	AFTER	COMMAND	DESCRIPTION
		DIMENSION BASELINE	Linear dimensions continued from a common baseline with the first extension line common to each dimension.
	—2.50—	DIMENSION LINEAR	An automatic command that will provide the dimension of an entity or segment once it has been identified with the cursor.
	R.50 Ø.75	DIMENSION RADIAL	Circles or arcs are automatically dimensioned after being selected.
ARROW SIZE .25 TEXT HEIGHT .18 SUPPRESS 1ST EXTENSION NO SUPPRESS 2ND EXTENSION NO TEXT ABOVE DIM. LINE NO	ARROW SIZE .18 TEXT HEIGHT .12 SUPPRESS 1ST EXTENSION YES SUPPRESS 2ND EXTENSION NO TEXT ABOVE DIM. LINE YES	DIMENSION VARIABLES	The values which determine the manner in which dimensions are drawn may be varied.
	8.6 5.64 45 4.2	DISTANCE / ANGLE (LIST, ID)	Describes spatial characteristics of a segment or between two points. The characteristics may include: distance between endpoints, angle, the change in X distance, the change in Y distance, and location with respect to the origin.
		DIVIDE (MEASURE)	Any entity may may be divided into a specified number of equal length intervals.
		DONUT	A solid circle, with or without a hole, may be automatically created.
		DRAG	Dragging enables a CAD drafter to visually move an object into place. Each time the cursor is moved (dragged), the object will be redisplayed at the new position.
		EDIT–LINE (STRETCH, EXTEND MODIFY, TRIM)	Any line or vertex may be altered. Such modifications include adding or deleting a vertex; joining, moving, stretching, extending, or shortening a line.
		ELLIPSE (ISOPLANE)	An ellipse may be automatically created. It may also be constructed to fit any of the three standard isometric planes or faces.

BEFORE	AFTER	COMMAND	DESCRIPTION
	MAIN MENU	END (QUIT)	When finishing work, the system must be deactivated and returned to the main menu. Be careful to first save the drawing if it is to be utilized further.
		ERASE	A command that will erase an object or an identified rectangular region.
		ERASE LAST	Keying this will cancel the last command entered.
		EXPLODE	Multiple entities blocked into a single object, as well as dimensions and polylines, may later be broken apart. Individual entities may then be altered without affecting the others.
		FILING (FILES IN/OUT)	After a drawing is complete, it may be permanently stored. A common filing procedure involves "off-loading" the contents magnetically onto a floppy disk.
		FILLET	A command that will draw an arc tangent to two selected lines at a specified radius.
HYDROGEN P & D	HYDROGEN P & ID	FIT	Text may be created so that it fits between any two points without affecting its height.
COMMAND LAYER	LAYER COLOR LINE-TYPE 1 WHITE CONTIN 2 BLUE DASH 3 RED CENTER	FLIP SCREEN (GRAPH SCREEN)	A single-screen system may be switched back and forth between the text and graphics display.
		GAP (BREAK P EDIT)	Any portion of a line, polyline, or segment may be altered.
DEFINE GRID		GRID	A network of equally spaced grid points that default to one unit of scale. Grids may be changed by keying in the horizontal and vertical spacing.

BEFORE	AFTER	COMMAND	DESCRIPTION
		HATCH (SECTIONING, AREA FILL, SOLID)	A command that will cross hatch or fill an area within an identified boundary.
HELP: ZOOM	ZOOM NUMBER ... ZOOM ALL ... ZOOM CENTER ... ZOOM WINDOW ...	HELP (MENU)	The HELP option provides specific information regarding each command. It may be accessed to explain each command option or define the method of execution.
		INSERT	A command that will scale up or scale down the size of an object or block by identifying it and specifying the scale factor.
	QTY. DESCRIPTION / 3 / 3 IN. GATE VALVE	ITEM LIST (BILL OF MATERIAL, ATTRIBUTE, DFX IN/OUT)	Text information assigned to an object may be stored or retrieved in a printed form. This is accomplished using attributes (DFX IN/OUT files.)
LEFT JUSTIFY RIGHT JUSTIFY	LEFT JUSTIFY RIGHT JUSTIFY	JUSTIFY	Rows of text may be "lined up" (justified) per a specific requirement. Left justify, for example, will line the starting point for each row of text with respect to an X coordinate location (vertical line).
		LAYER (LEVEL)	A layer or plane on which a drawing or a portion of a drawing is located. Layers may be "switched" on or off, allowing various combinations of a drawing to be displayed.
		LEADER	The automatic creation of an arrowhead located at the end of two or more segment lines.
	⊕ .006 A	LIBRARY	A collection of stored groups of symbols that may be called up for use on a drawing.
		LINE (SOLID)	A command that will allow the creation of straight lines from point to point. (*Also see* polar, absolute, and relative coordinates.)
		LINE TANGENT	A command that will automatically draw a line tangent to two circles or arcs after the two curved surfaces are selected.

BEFORE	AFTER	COMMAND	DESCRIPTION
		LINE THICKNESS (TRACE)	A command that will widen a line or a curve after the desired width has been keyed into the system.
		LINE TYPE	Standard line types such as hidden, center, and phantom may be created. The length of each segment may also be varied to suit particular applications.
		MIRROR	A process that results in a hinged, reflected, or mirrored image around a user specified axis.
		MODEL	3-D models automatically display isometric or perspective views either in wireframe (edge views) or solid (shaded) while concealing hidden lines.
		MOVE	A command that will move an object or group of objects to any location.
		MOVE NEXT POINT	A gap between points while in the LINE command.
		MULTI-SEGMENT LINE	The creation of continuous line segments. The result is considered a single line comprised of multiple segments.
		OBJECT SNAP (OSNAP)	A command that locks points to various locations on an object. Common positions include endpoints, midpoints, tangent, center, nearest, quadrant, and perpendicular.
		ORTHO-ON (HORIZONTAL OR VERTICAL)	ORTHO automatically forces a line to be drawn on a horizontal or vertical axis only.
		ORTHO-OFF	A command that allows the creation of diagonal lines at any angle.

BEFORE	AFTER	COMMAND	DESCRIPTION
		PAN	A scan to a different portion of a drawing while zoomed in.
		PARALLEL (OFFSET)	Lines may be created from an existing line or curve after selecting the command and keying in the offset distance.
PEN 1	PEN 2	PEN	Pen numbers refer to the plotting of a drawing. A number 2 pen selection will be plotted with a number 2 pen.
	PERIMETER=11.50	PERIMETER	A command that will calculate and display the perimeter distance around an object.
		PERPENDICULAR LINE	An option that will create a line at 90 degrees (perpendicular) to another.
		PLINE (SPLINE)	A command that will construct an irregular curve after points along that curve have been identified.
		PLOT	A pen plotter will create a finished drawing. Linework and lettering of professional quality will be produced.
		POINT	Points may be located any place on the screen by either keying in the coordinate or picking a curser position.
		POLAR COORDINATE	A second point coordinate is located by keying its distance and angle from the first point.
		POLYGON	A polygon of any number of sides (greater than 2) may be created to a specified radius.

BEFORE	AFTER	COMMAND	DESCRIPTION
		POLYLINE	A command that results in a series of connecting lines between cursor points (considered a single object).
		RECTANGLE	Create a rectangle by picking diagonal corners with the cursor or keying in the coordinates.
		REDRAW	After deleting, the display often will appear messy. Redrawing will redisplay the objects on the screen, "cleaning up" the display.
59%	28%	REGENERATE (PACK)	Accumulated revisions and useless revision work utilizes a significant amount of drawing storage space. Data packing will discard this from drawing memory, leaving more room for useful work.
L	6,5 L	RELATIVE COORDINATE	A coordinate point located at an X and Y distance from (relative to) the last point digitized.
		ROTATE	An object or group of objects may rotate about an origin at any specified angle.
		RUBBERBAND	As the cursor moves across the screen, the entity (line, circle, arc) will be redisplayed. This provides an excellent visual guide for optimizing the entity size and position.
	COMMAND: SAVE NAME: YOKE	SAVE	It is desirable to periodically update the drawing file to disk. This way in the event of a power failure, or lock up, all work will not be lost.
		SCALE	A command that will scale up or down the size of an object or group of objects when specifying any scale factor.
LIMITS X:0,11.00 Y:0,8.50 SCALE: FULL UNITS: DECIMAL	LIMITS X:0,34.00 Y:0,22.00 SCALE: HALF UNITS: DECIMAL	SETUP (LIMITS, SCALE, UNITS)	When beginning a new drawing the parameters may be selected.

BEFORE	AFTER	COMMAND	DESCRIPTION
		SKETCH	This command permits freehand drawings to be created. It is used for irregular application such as map outlines.
		SNAP-ON	A mode that will automatically snap the cursor to the nearest snap increment. The snap increment is often set to the grid spacing or half the grid spacing.
		SNAP-OFF	A mode that will allow any point to be located, regardless of the grids or snap increment.
	LIMITS X:0,11.00 Y:0,8.50 SCALE: FULL UNITS: DECIMAL	STATUS (SETVAR)	The established parameters for a particular drawing may be displayed on the monitor.
		STRETCH	Moves a selected portion of a drawing to a new position while retaining the original form.
STANDARD FONT	Italicized font	TEXT-STYLE (FONT, SCRIPT)	Varios text style options are available and may be selected in lieu of the default standard font.
TEXT PLACEMENT	TEXT PLACEMENT	TEXT (NOTES)	Text is created (line by line) on the keyboard and transferred to the drawing by cursor position select.
⊢2.50	⊢2.50	TOLERANCE	Plus and minus limits may be independently specified. The tolerance will automatically be added next to the segment dimensioned.
		TRACE	A command that will draw a single line as a double line after the desired width has been keyed into the system. The spacing between the lines may be filled in solid or left open.
		TRIM (CLIP)	A command that will automatically trim or clip an object back to another object.

BEFORE	AFTER	COMMAND	DESCRIPTION
		UNDO (OOPS, REDO)	This option is handy to use to immediately restore any portion of the drawing that had inadvertently been erased.
		Z AXIS	3-D modeling coordinate points are located at a horizontal distance (X), a vertical distance (Y), and a depth distance (Z), from a O,O,O point (drawing origin).
		ZOOM	Magnification of a certain portion of the drawing.
		ZOOM ALL (FULL WINDOW)	A command that will display the complete drawing on the monitor.

WELDING

DESIGN THEORY

Before we begin machine element design, the fundamentals of mechanical support will be presented. Mechanical supports include welded assemblies and chassis patterns. These can be categorized as thick-wall (weldments) and thin-wall (sheet-metal layout) mechanical applications. Welded assemblies are helpful in design project work. Parts or assemblies holding machine elements may have to be fastened permanently. Any casing or housing using steel plate fabrication could potentially apply weld techniques. This chapter will cover the ways to accomplish this. The knowledge gained will be useful in later chapters when larger projects will be developed.

All welding drawings will be CADD-generated. The application will be an extension of introductory CADD topics. If you have completed an introduction to CADD course, you should have little difficulty creating each weld drawing. The only new concept includes the use of a weld symbol library. If one is not available, you may utilize the concepts learned in Introduction to CADD. That is, create each symbol, group or block it into a subassembly, and store it for future use. If you decide not to create a symbol library, each may be created using the basic line, leader, and text commands.

Standard Symbols

The basic weld symbol is shown below:

How it is labeled depends upon the type of weld used. Each type is signified by a different symbol.

Only the most common will be covered in this chapter. These include:

TYPE	SYMBOL	TYPE OF JOINT	
Fillet			Parts joined at an angle
Butt			Medium thickness
Groove			Thick plate
Spot			Thin plate

The left of each symbol is normally reserved for specifying the weld size. Other information is also presented on the symbol. This is best shown by the example in Fig. 1-1. It is read as follows:

.25 in. is the size (leg length)

Fillet weld (this side—symbol below line)

4.00 in. long

Done in the field (flag at the vertex)

Four different locations (tail note)

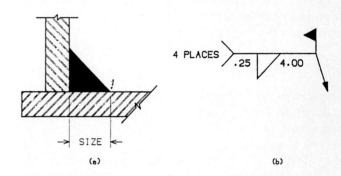

Fig. 1-1 (a and b). Basic weld identification.

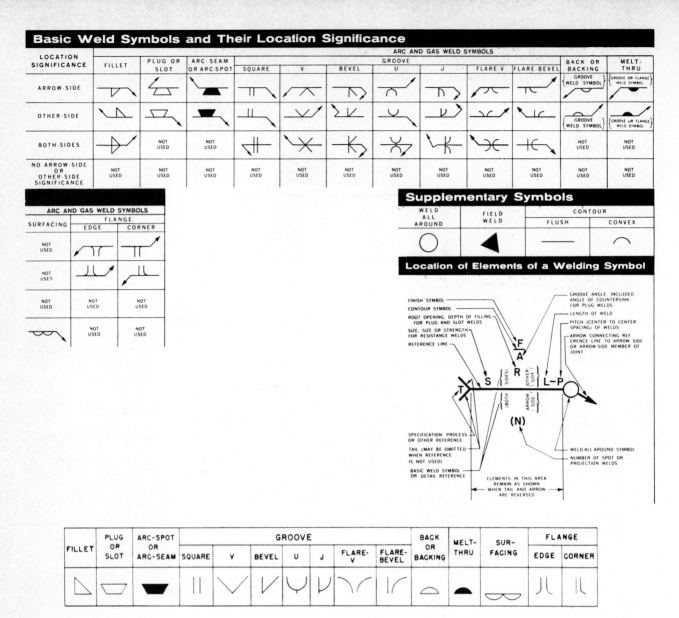

Fig. 1-2. *Standard weld symbols (Lincoln Electric).*

This text will not elaborate on the ways to interpret every type of weld. There is a wealth of information to be found in virtually every traditional design or drafting text. Refer to the standard weld symbol chart in Fig. 1-2, and review it. To assist you in reviewing, you may also refer to the illustrative examples shown in Fig. 1-3.

Weld Strengths

After the fundamentals of welding drawings are understood, the next step is to apply the principles to design projects. This first involves determining the type of weld to be used, and second, the size of weld to specify. The types of applications are unlimited. Primarily, however, fillet and groove are preferred in this text.

FILLET

The weld strength of fillet-welded plate depends upon the size of the weld. This is measured by considering the weld length as a side of a 45° right triangle as shown in Fig. 1-1(a). The relationship between weld size and plate thickness for a "minimum-strength" weld is:

MINIMUM FILLET WELD SIZE (INCHES)	PLATE THICKNESS (INCHES)
.12	.25
.19	.50
.25	.75
.31	1.25
.38	2.00
.50	6.00
.62	Over 6.00

Courtesy: American Institute for Steel Construction (AISC).

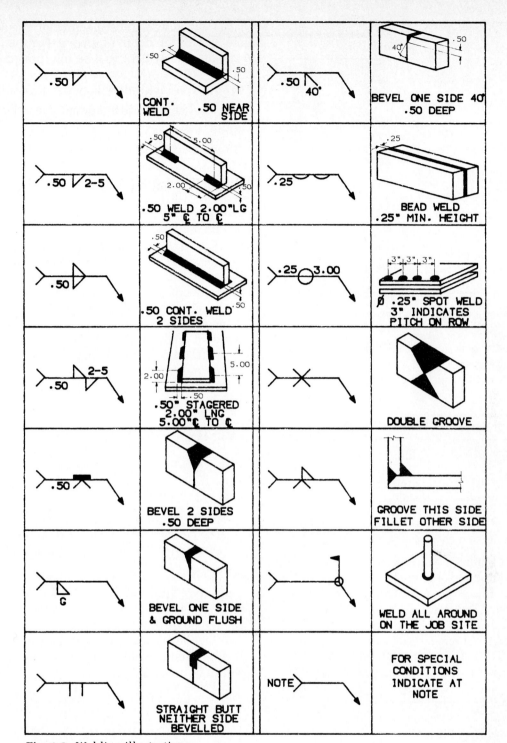

Fig. 1-3. Welding illustrations.

This relationship refers to the minimum fillet weld to be used for sufficient rigidity. It is considered to be 33 percent of a full-strength weld. If the parts to be welded fall between these values, select the next largest.

The maximum fillet size, which is referred to as the "full-strength" weld, is .75 times the plate thickness of the smaller section to be welded. Often the two sections will vary in size. When designing, strive to make each section size as close as possible with the weld size never exceeding the plate thickness. Standard steel plate stock sizes are shown in Fig. 1-4.

BUTT AND GROOVE

When plates must be permanently joined and meet in parallel, groove or butt welds generally will be used. The type of weld and surface preparation at the end of the plates depends on plate thickness. The relation between type of weld, weld size, and plate

SIZE (INCHES)	WEIGHT (PER FT.^{2})
.188	7.66
.250	10.21
.312	12.76
.375	15.32
.438	17.87
.500	20.42
.562	22.97
.625	25.53
.750	30.63
.875	35.74
1.000	40.84
1.125	45.95
1.250	51.05
1.375	56.16
1.500	61.26
1.625	66.37
1.750	71.47
1.875	76.58
2.000	81.68

Fig. 1-4. Standard steel plate thicknesses.

thickness, where practical, shall conform to the requirements shown in Fig. 1-5.

Use the following type of weld for different section thicknesses:

THICKNESS (INCHES)	TYPE OF WELD
Less than .12	Spot
.12 to .25	Butt
.25 to .50	Single groove (V-double bevel)*
.50 to 1.00	Double groove (V-double bevel)
Over 1.00	Double J

*A bevel is similar to a chamfer (see illustration).

If bending of the plates is involved, maximum plate thickness for double-sided groove welds is 1 in.

Notice from Fig. 1-5 that the distance W between each plate during the welding process will vary. For a single-sided groove weld joining two .25-in. plates, butt the plates together. For the same weld joining .50-in. plates, leave a spacing of .19 in. For plates between this range, vary the spacing proportionally. The

spacing, if any, will be specified on the weld symbol along with the depth of groove. For the .50-in. plate example the symbol will be interpreted as follows:

Groove weld with double bevel (this side)

60° bevel angle (this is normally standard)

.38-in. bevel depth (T − F, Fig. 1-5)

.19-in. spacing between plates (W)

Plate Thickness

It may be necessary to convert a casting into a steel weldment. Steel has greater strength and rigidity than cast iron. Thus a smaller thickness may be used. Remember, however, when converting, be certain you are not changing how the part functions. Also, do not vary any critical dimensions, such as hole locations.

One classification of gray cast iron is the American Society for Testing and Materials (ASTM) series. The standard grades are ASTM 20, 25, 30, 40, 50, and 60. A typical grade 30 yields approximately a 2:1 strength ratio. Thus a 1.00-in.-thick iron section may be substituted for a .50-in. steel plate.

DESIGN APPLICATION

Fillet Welds

SPECIFICATION

Two 1.00-in.-thick plates that meet at right angles must be permanently joined. Determine the minimum-strength and full-strength weld conditions.

SOLUTION

Since the plates are to be joined at an angle, a fillet weld should be utilized. Refer to the minimum weld strength table in the Section "Weld Strengths." Since 1.00-in. plate is not listed, select the next largest, 1.25. The minimum weld strength is .31. Thus the weld would be specified as:

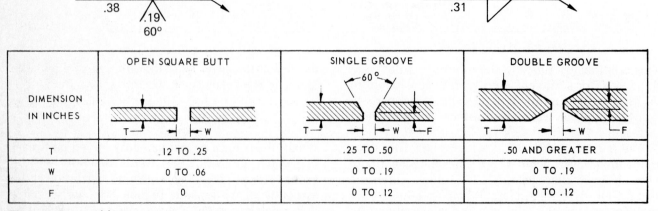

DIMENSION IN INCHES	OPEN SQUARE BUTT	SINGLE GROOVE	DOUBLE GROOVE
T	.12 TO .25	.25 TO .50	.50 AND GREATER
W	0 TO .06	0 TO .19	0 TO .19
F	0	0 TO .12	0 TO .12

Fig. 1-5. Butt weld sizes.

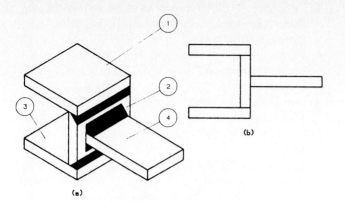

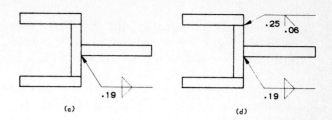

Fig. 1-6 (a–d). *Weldment.*

If the weld is to be full strength, the size will be .75 times the thinnest section. This weld would be specified as:

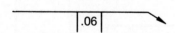

Butt Welds

SPECIFICATION

Two parallel .25-in.-thick plates must be permanently joined. Determine the type and size of weld.

SOLUTION

Refer to Fig. 1-5. A .25-in. plate falls within both the open square butt and single-groove categories. Either may be used. It is the maximum recommended thickness for a butt weld. Thus the largest W dimension should be used. The weld would be specified as:

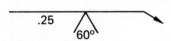

If you desire to use a single-groove, double-bevel weld, .25 in. is the smallest recommended thickness. In this event use the smallest W and F dimensions listed (both are zero). The weld would be specified as:

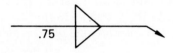

Casting to Weldment

SPECIFICATION

Convert a 1.00-in.-thick, ASTM grade 30 cast-iron section to a full-strength fillet weld assembly.

SOLUTION

1. The strength ratio is 2:1; thus .50-in. steel plates may be used.

2. The full-strength weld size is 75 percent of the thinner section to be joined; .38 (.50 × .75) fillet welds are necessary for a full-strength condition.

CADD DRAWING

Create a minimum strength weld assembly for the drawing shown in Fig. 1-6(a). Plates 1, 2, and 3 are .38 plate × 4 × 4 in. long. Plate 4 is .38 plate × 3 × 4.25 in. long. Prepare the drawing as follows:

1. Activate the CADD system.
2. Create the drawing file, and SETUP the units (B size and full scale).
3. Determine the type and size of welds. A fillet weld must be used to join part 4 to part 2. For .38-in. plate, use a .50-in. plate in the table from the Section "Weld Strengths." Select a minimum-strength weld of .19 in. Use groove welds to join parts 1 to 2 and 2 to 3. Since one of the plates does not have sufficient thickness, they should be single bevel as shown. Using Fig. 1-5, we will find the depth of groove to be .25 in. (T − W = .38 − .12) and the spacing .06.
4. Select the LINE command.
5. Use SOLID lines to crate the front view as shown in Fig. 1-6(b).
6. Select the WELD symbol library.*
7. Select FILLET.*
8. Place the fillet weld as shown on Fig. 1-6(c). Remember to add the size.*
9. Select GROOVE-SINGLE BEVEL.*
10. Place the groove weld as shown on Fig. 1-6(d). Remember to add the size information.*
11. Add a top or profile view as desired using LINE and HIDDEN LINE (or LAYER).
12. Use TEXT to specify the size and material of each part.

---

*Note: If a weld symbol library is not available, you may create your own symbols and store them. If this is not desirable, create each symbol using LEADER, LINE, and TEXT.

ASSIGNMENTS

1. Complete the drawing to join the three parts shown in Fig. 1-7 by specifying the correct weld symbols. Use minimum-strength welds.

2. Prepare full-strength weld working drawings for the objects shown in Fig. 1-8. Use the following sizes:

IDENTIFIER	PART (a)	PART (b)
A	φ.25—2 required	φ #12—2 required 3-in. C-C
B	2.50	R1.25
C	1.25	R1.00
D	.62	R.19
E	.25	.50
F	.75	1.38
G	.38	1.50
H	R.25	
J	.50	

3. Prepare a multiview welding drawing for thebracket shown in Fig. 1-9(a). Use minimum-strength fillet welds.

4. Prepare a multiview welding drawing for the bracket shown in Fig. 1-9(b). Use full-strength fillet welds.

5. Prepare a multiview welding drawing for the base plate shown in Fig. 1-10. Use either minimum- or full-strength welds as assigned by your instructor.

6. Convert the ASTM grade 30 casting shown in Fig. 1-11 into a welding drawing. Use full-strength welds for the joints requiring a fillet.

7. Prepare a welding drawing of the fan and motor base assembly shown in Fig. 1-12. Use minimum- or full-strength welds as assigned.

8. Prepare a welding assembly drawing of the reservoir shown in Fig. 1-13. The reservoir must be capable of storing at least 25 gallons of fluid (1 gal = 231 in.^{3}). Use a .38-in. plate. Select the appropriate type and size of all welds.

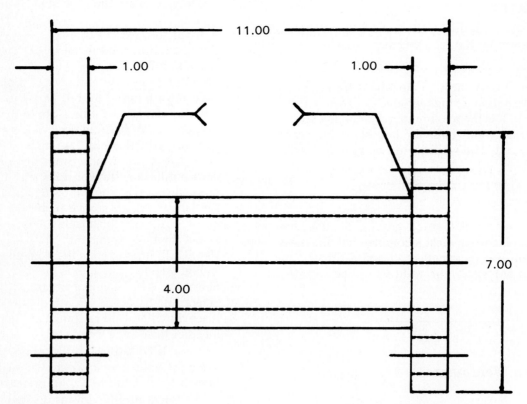

Fig. 1-7. Weldment.

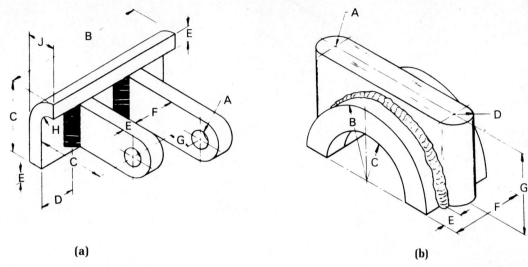

(a)

(b)

Fig. 1-8 (a and b). Weld drawings.

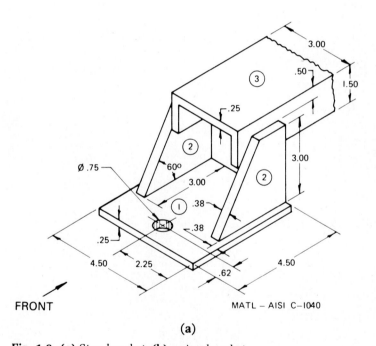

FRONT

MATL – AISI C–1040

(a)

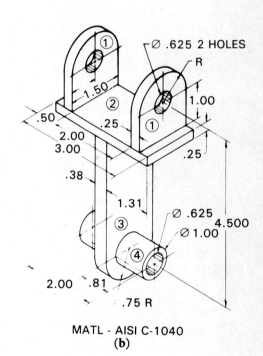

MATL - AISI C-1040
(b)

Fig. 1-9. **(a)** Step bracket; **(b)** swing bracket.

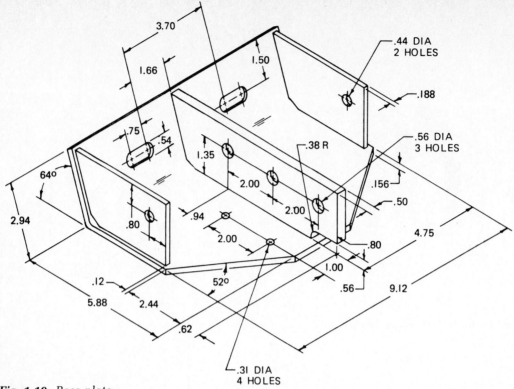

Fig. 1-10. Base plate.

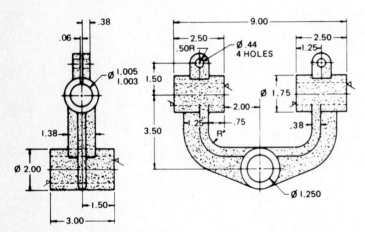

Fig. 1-11. Connecting bracket.

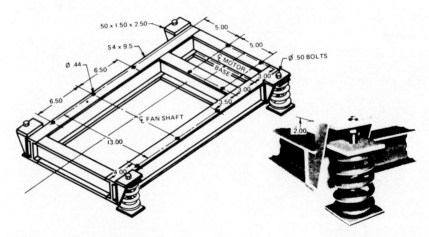

Fig. 1-12. Fan and motor base.

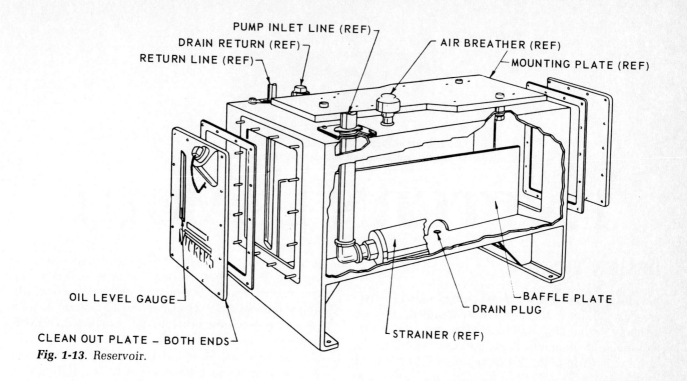

PUMP INLET LINE (REF)
DRAIN RETURN (REF)
RETURN LINE (REF)
AIR BREATHER (REF)
MOUNTING PLATE (REF)

OIL LEVEL GAUGE

BAFFLE PLATE
DRAIN PLUG

CLEAN OUT PLATE — BOTH ENDS
STRAINER (REF)

Fig. 1-13. Reservoir.

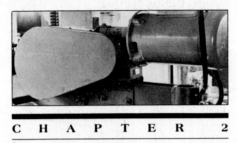

C H A P T E R 2

PATTERN LAYOUT

DESIGN THEORY

Besides welding, a sheet-metal pattern may be a necessary detail for a mechanical design project. Perhaps a chassis, or housing, will be used to support and protect machine elements. Also, brackets are handy for holding parts in place. This chapter will provide the information you will need for such applications as they may arise.

Thin Gage

A sheet-metal pattern development is a housing that is laid (stretched) out as a flat surface. It is obtained by unfolding or unrolling the sheet-metal part. Every line, bend, and area will be seen true size. This is similar in concept to the sheet of paper you are reading. If you look perpendicular to the page, the true shape is seen. You can measure its true width, height, and so on. With a sheet-metal pattern, the fabricator can see how much metal is required and perhaps even use it as a template. Other instructions will also be provided on the layout so that the part may be bent into its desired shape. A typical flat pattern layout is illustrated in Fig. 2-1. Notice that the part is also shown in its finished (final configuration) form.

A working pattern must include all of the following:

1. **Pattern Layout.** Apply geometric methods, such as true length and size theory, as necessary.

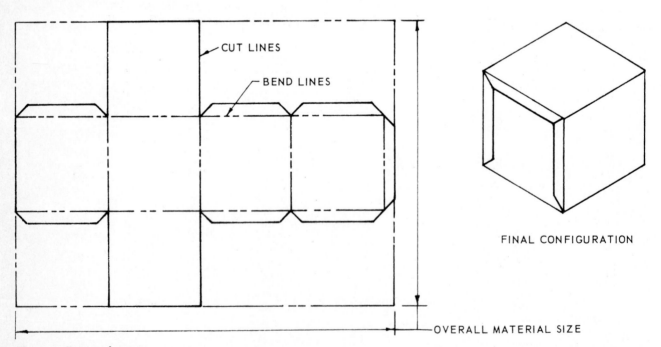

Fig. 2-1. *Pattern layout.*

MECHANICAL DESIGN USING CADD

10

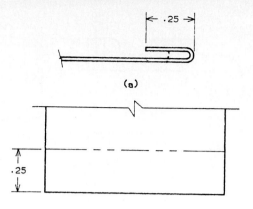

(a)

(b)

Fig. 2-2. (a) Single hem; (b) single hem layout.

2. **Overall Material Size.** Determine the smallest rectangular area required to fabricate the pattern.

3. **Dimensions.** Include all dimensions necessary to lay out the part.

4. **Bending Notes and Identification.** Indicate each bend line with a phantom, lighter line or with some other identification (x, o). Identify, using a leader line or note, the direction and degree of bend.

5. **Seams.** Specify the desired types of seams, edges, or joints. For sheet-metal application in this text, a .25-in., single-hem (180° bend) edge seam will be used. This is illustrated in Fig. 2-2.

6. **Type of Material.** Specify the gage and type of material with the overall material size. Standard sheet-metal gages are listed in Fig. 2-3.

7. **Final Configuration.** Include pictorial or ortho-

graphic views of the part. As in Fig. 2-1, it will show how it appears after all bending is complete.

8. **Tolerance.** Under normal sheet-metal applications, the tolerance allowance is fairly loose. Some of the applications even trim to fit. The exact tolerance, however, will depend upon the specific application.

9. **Notes.** Complete the working drawing with applicable notes. A note commonly used will indicate that sharp corners are not allowed.

Thick Gage

Suppose you have a straight piece of metal and wish to bend it. Figure 2-4(a) illustrates a part that has a 90° bend. How much straight material is used for this bend?

For thin-gage with small bend radius applications, the amount of material required is not considered since it is small. This is especially true when the application is accompanied by a loose tolerance. For thicker-gage steel, however, the amount of material becomes significant. It must be considered. The distance around the bend is known as the *bend allowance*. To determine it, compute the following equation:

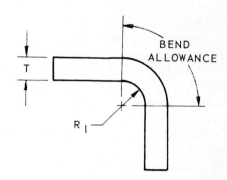

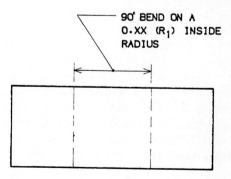

Fig. 2-4 (a and b). Developed length.

GAGE	THICKNESS	
	COLD ROLLED SHEET (IN.)	GALVANIZED SHEET (IN.)
10	.134	.138
11	.120	.123
12	.105	.108
13	.090	—
14	.075	.078
15	.067	—
16	.060	.064
17	.054	—
18	.048	.052
19	.042	—
20	.036	.040
21	.033	.037
22	.030	.034
23	.027	.031
24	.024	.028
25	.021	.025
26	.018	.022
27	.016	—
28	.015	.019
30	—	.016

Fig. 2-3. Standard U.S. gages.

$$BA = (.71t + 1.57 \, Ri) \times N/90 \qquad \text{2-1}$$

where BA = bend allowance (in.)*

.71 = constant to allow for neutral bend axis, which varies somewhat with thickness

t = thickness of the sheet (in.)*

1.57 = constant equal to one-half of π (3.14)

Ri = inside radius of bend (in.)*

N = number of degrees of bend

90 = constant

Casting

At certain times, conditions will not allow for the use of a sheet-metal housing. A gear reducer, for example, may require constant lubrication. It would be better to use a casting for an application of this nature. This way the gears may continuously function while submerged in an oil bath. It is not the intention of this text to provide all information regarding the casting process, which can be a course in itself. As this condition may arise, however, simply determine your critical dimensions. This includes distances between centers of shafts, clearances, casting wall, and boss thicknesses. These are referred to as dimensions A, B, C, and D, respectively, in Fig. 2-5(a).

A. The distance between shaft centers will be determined by the mechanical component size (shown hidden).
B. Be sure to leave some clearance so that the mechanical components will not scrape casting walls. One-quarter to one-half (.25–.50) in. will be enough.
C. Make the casting wall .38 in. thick. This will account for flaws that are commonly found in castings.
D. The boss will be extended so that a bearing and seal will fit in the casting. To obtain this dimension, add the width of each component. Be generous with the size since this is the part of the casting that supports the loads—but more about that later when you will learn to size bearings.

When the casting detail is actually prepared (by others) attention will have to be given to generous radii, fillets, and rounds. Notice the large radii provided at the four corners. This will reduce the likelihood of cracking, and consequently failure, at these higher-stressed areas. All cast corners are rounded. Thus a command FILLET is important for these types of drawings. It will assist the creation of each view more

---

*Refer to Fig. 2-4. Since 90° and 180° bends are common, the above formula is set up for ease of mathematical manipulation.

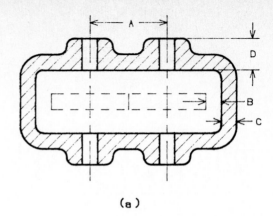

(a)

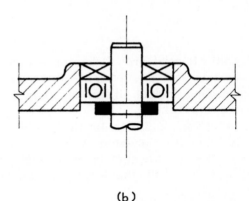

(b)

Fig. 2-5. (a) *Casting dimensions;* (b) *ZOOM in—seal, bearing, and retaining ring.*

accurately and rapidly. Generally, an overall note such as the following will be given to specify fillets and rounds:

All fillets and rounds to be R.25 unless otherwise specified.

Zooming in on one of the bosses shows a seal, bearing, and retaining ring assembled on the shaft and casting as illustrated in Fig. 2-5(b).

DESIGN APPLICATION

Thin Wall

SPECIFICATION

Design a container for the product illustrated in Fig. 2-6.

SOLUTION

Once the size of the container product is known, the pattern may be determined. Since this is a flat-wall

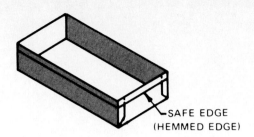

—SAFE EDGE
(HEMMED EDGE)

Fig. 2-6. Isometric.

application, the design is easy. Every surface and bend line is true size. Thus the following surfaces will be required:

One bottom—8 × 12 in. long
Two sides—3 × 12 in. long
Two sides—3 × 8 in. long
A .25-in. single hem at the top of each side piece

Thick Wall

SPECIFICATION

A .25-in. steel plate is to be bent 90° at a .50-in. inside radius. It will have two straight portions extending .75 and 1.00 from each side of the bend. The finished piece will appear similar to the one in Fig. 2-4(a). What is the total developed length?

SOLUTION

Step 1
Determine the bend allowance by solving Eq. 2-1.

$$BA = (.71t + 1.57\ Ri) \times 90/N$$

$$= [(.71 \times .25) + (1.57 \times .50)] \times 90/90$$

$$= (.177 + .785)\ 1 = .962 = .96\ in.$$

Step 2
The total developed length becomes

Straight sections + bend allowance

$$.75 + 1.00 + .96 = 2.71\ in.$$

Step 3
The pattern layout will be similar to that shown in Fig. 2-4(b). Instead of XX in. inside radius, however, specify .50 in.

CADD DRAWING

A sheet-metal pattern drawing of the part illustrated in Fig. 2-6 may be prepared as follows:

1. Boot the system and SETUP a C size FULL scale drawing.
2. Select GRIDS, and set the increment spacing to 1.00 in. with the SNAP ON.
3. Select LINE and PHANTOM (or PEN 2).
4. Draw four lines to form the bottom of the container. The two horizontal lines are 12.00 in. long, and the two vertical lines are 8.00 in. long. It should appear as shown in Fig. 2-7(a).
5. Create each long sidewall by drawing a 12.00-in. horizontal line as shown in Fig. 2-7(b).
6. Create each short side wall by drawing an 8.00-in. vertical line as shown in Fig. 2-7(c).
7. ZOOM in on the upper long side.
8. Select GRIDS, and set the increment spacing to .25 in. with the SNAP ON.
9. Select SOLID LINE and PEN 1.
10. Create the single hem by drawing a vertical line up 3.25, a horizontal 12.00, and a vertical down 3.25 as shown in Fig. 2-7(d).
11. PAN to another side. Repeat Step 10 using the appropriate length for that side (8.00 or 12.00 in.).
12. Repeat Step 11 for the remaining sides.
13. Remove the grids by selecting GRIDS OFF.
14. Return the drawing to its original size by selecting ZOOM ALL. It will appear as shown in Fig. 2-7(e).
15. Select LEADER and NOTE.
16. Add the bend information as shown in Fig. 2-7(f).
17. Select DIMENSION-LINEAR (HORIZ AND VERT).
18. Dimension the layout as shown in Fig. 2-7(g).
19. Select TEXT and .125 lettering height.
20. Add the notes as shown in Fig. 2-7(h).
21. If your system has a modeling option, it is possible to automatically generate a 2.5-D wireframe isometric of the finished part. Key in the appropriate height dimensions before selecting VIEW-POINT and HIDE options. The isometric with hidden lines removed will appear. If your system does not have a modeling option, create the final configuration using an isometric (30° angle) grid pattern.

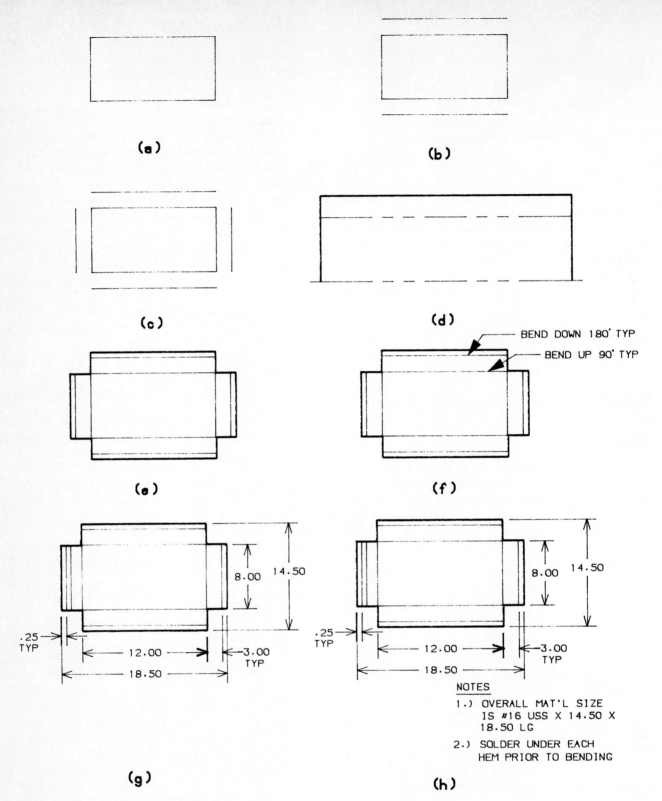

(a)

(b)

(c)

(d)

BEND DOWN 180° TYP

BEND UP 90° TYP

(e)

(f)

(g)

8.00

14.50

.25
TYP

12.00

3.00
TYP

18.50

(h)

8.00

14.50

.25
TYP

12.00

3.00
TYP

18.50

NOTES

1.) OVERALL MAT'L SIZE
IS #16 USS X 14.50 X
18.50 LG

2.) SOLDER UNDER EACH
HEM PRIOR TO BENDING

Fig. 2-7 (a–h). Working pattern drawing.

ASSIGNMENTS

1. Prepare a sheet-metal working drawing for a container 3.50 in. high × 6.00 in. wide × 11.00 in. long. Use #25 gage galvanized steel and a .25 single hem for each edge. Follow the procedure used in the Section "CADD Drawing."

2. Prepare a sheet-metal layout of a transition piece as shown in Fig. 2-8. It is to symmetrically join a 4 × 6 rectangular duct to a 6 × 10 rectangular duct. Use #30 gage galvanized steel. Add a note to "break all sharp edges."

3. Prepare a pattern drawing for the container shown in Fig. 2-9. Provide bend lines and an opening tab. Score lines for the .25-in.-wide opening tab should be specified to leave 55 to 60 percent of the original material thickness. Use #30 gage material. Specify that corner seams are required to fasten the top and bottom. Also specify that each seam is to be soldered during assembly. Do not concern yourself about how the corner seam will look or how it is made. Just leave .50-in. material all around the top and bottom pieces and .25-in. material to attach to each side of the long strip. Use the bend allowance formula to determine the developed length of the strip. Be sure to include all items necessary for a working drawing as outlined in the Section "Design Application."

4. Determine the developed length required to manufacture the 1.50-in.-wide bracket shown in Fig. 2-10. The top hole centrally located has a .50-in. diameter. The bottom hole has a centrally located .38-in.-diameter NC thread. Prepare a working drawing of the part.

5. Prepare a pattern drawing of a rectangular to round transition piece. It is to join a 12.00-in. square to an 8.00-in. diameter and is 10.00 in. high as shown in Fig. 2-11. Use #30 gage steel. *Note:* Do not attempt to solve this problem unless you have a descriptive geometry background.

6. Prepare a sheet-metal housing for the fan assembly shown in Fig. 3-8. Reshape the housing for a 12.00-in-diameter × 6.00-in.-wide fan. Provide 4.00- × 8.00-in. openings for rectangular HVAC ducts. Each shaft hole has a 1.00-in. diameter. Leave 2.00 in. as clearance between the fan and either side of the housing. Thus the housing will be 10.00 in. wide.

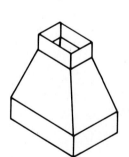

Fig. 2-8. Transition piece—
rectangular to
rectangular.

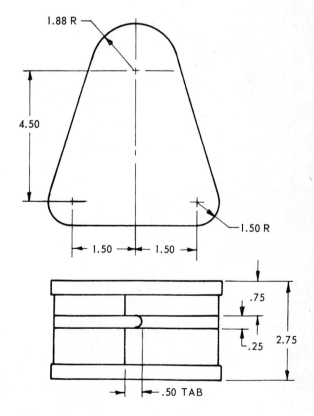

Fig. 2-9. Container.

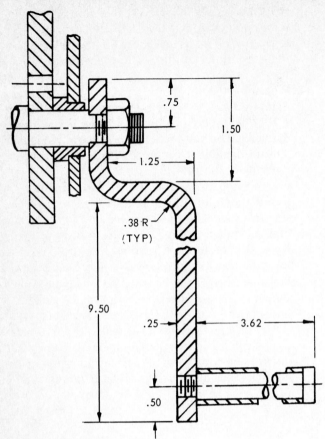

Fig. 2-10. Winch handle.

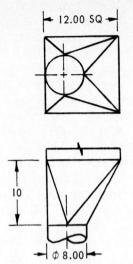

Fig. 2-11. Transition piece—square to round.

C H A P T E R   3

BELT DRIVES

DESIGN THEORY

Belt drives are used in many applications to transmit power and motion. There are several different types of drives, including:

- Flat belts
- V-belts
- Grooved (Poly V)
- Synchronous (timing)

V-belts, most commonly used for low- to medium-horsepower applications, will be analyzed in this chapter. The power is transmitted between a belt and a pulley as shown in Fig. 3-1. In general, the more power that must be transmitted, the larger the belt. Cross-sectional areas of standard size industrial belts are shown in Fig. 3-2. The belt size is a function of the horsepower rating table shown in Fig. 3-3(a). If the horsepower at a particular speed is known, the belt size (O, A, or B) and minimum-required pulley diameter may be determined using this table. For higher-horsepower ratings that exceed table values, multiple strand belts will be used. Thus if a single belt will safely transmit 1.50 hp, two belts will safely transmit 3.00, and so on. When we are specifying a belt drive system, the belt length must be considered. The length will be determined with the aid of the table shown in Fig. 3-3(b). The pulley diameters and center distance between pulleys are used for this purpose.

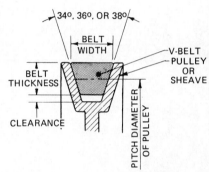

Fig. 3-1. V-belt and pulley.

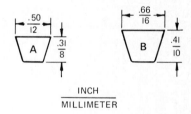

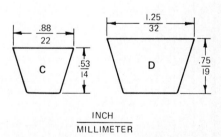

Fig. 3-2 (a–d). Industrial V-belts.

DESIGN APPLICATION

Single Strand

SPECIFICATION

A .50-hp, 1750-rpm motor (see Appendix 1 for standard motor sizes) is designed to operate a drill press.

It has an output spindle speed of approximately 1200 rpm. The center distance between the motor shaft and the spindle is approximately 20 in. Specify a drive system that will safely transmit this power.

HORSEPOWER RATINGS

RPM of small pulley	pitch diameter of small v-pulley—inches														
	1.50	1.75	2.00	2.25	2.50	2.75	3.00	3.25	3.50	3.75	4.00	4.25	4.50	4.75	5.00
200	. . .	. . .	. . .	0.08	0.12	0.15	0.18	0.22	0.25	. . .	0.18	0.22	0.24	0.28	0.29
400	. . .	. . .	0.06	0.08	0.12	0.18	0.22	0.27	0.32	0.31	0.35	0.42	0.46	0.52	0.56
600	0.04	0.07	0.08	0.12	0.18	0.22	0.27	0.32	0.36	0.44	0.51	0.58	0.66	0.73	0.81
800	0.05	0.08	0.11	0.15	0.22	0.28	0.34	0.41	0.45	0.55	0.64	0.74	0.81	0.93	1.00
1000	0.06	0.10	0.12	0.18	0.26	0.33	0.42	0.48	0.55	0.64	0.75	0.86	0.99	1.10	1.21
1160	0.07	0.11	0.15	0.21	0.29	0.38	0.46	0.54	0.62	0.69	0.84	0.98	1.07	1.23	1.35
1400	0.08	0.12	0.17	0.23	0.33	0.43	0.53	0.64	0.74	0.84	0.96	1.10	1.25	1.42	1.55
1600	0.08	0.14	0.19	0.25	0.36	0.48	0.58	0.69	0.80	0.90	1.02	1.20	1.36	1.53	1.68
1750	0.08	0.15	0.20	0.25	0.38	0.51	0.63	0.74	0.85	0.96	1.08	1.25	1.43	1.61	1.78
2000	0.09	0.16	0.22	0.28	0.41	0.55	0.68	0.81	0.92	1.05	1.17	1.35	1.54	1.73	1.90
2200	0.09	0.17	0.24	0.31	0.44	0.58	0.72	0.86	0.99	1.12	1.25	1.41	1.61	1.80	1.99
2400	0.10	0.18	0.25	0.32	0.45	0.61	0.76	0.91	1.05	1.19	1.32	1.45	1.65	1.86	2.02
2600	0.10	0.19	0.26	0.35	0.47	0.64	0.79	0.96	1.09	1.24	1.38	1.48	1.69	1.89	2.09
2800	0.11	0.19	0.28	0.36	0.48	0.66	0.83	0.99	1.14	1.28	1.42	1.48	1.71	1.91	2.11
3000	0.11	0.21	0.29	0.39	0.49	0.68	0.85	1.02	1.18	1.32	1.46	1.48	1.69	1.89	2.08
3200	0.11	0.21	0.30	0.39	0.51	0.70	0.88	1.05	1.20	1.36	1.50	1.50	1.67	1.86	2.03
3500	0.12	0.22	0.32	0.41	0.51	0.71	0.90	1.07	1.23	1.38	1.52	1.52	1.61	1.78	1.94
3600	0.12	0.22	0.33	0.42	0.52	0.72	0.91	1.09	1.25	1.40	1.54	1.54	1.54	1.71	1.85
3800	0.12	0.22	0.33	0.42	0.52	0.72	0.92	1.09	1.25	1.41	1.54	1.54	1.54	1.59	1.72
4000	0.12	0.22	0.34	0.44	0.53	0.72	0.92	1.10	1.26	1.40	1.52	1.52	1.52	1.52	1.55

For [hatched] Background use a — O — .38 wide, .22

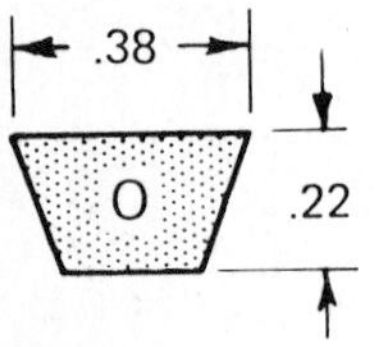

For [white] Background use a — A — .50 wide, .31

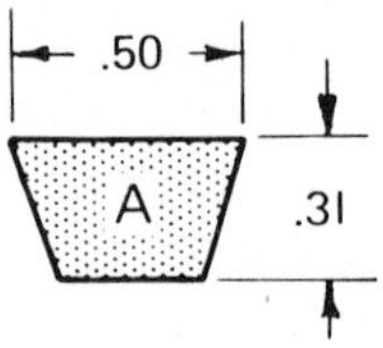

For [gray] Background use a — B — .66 wide, .41

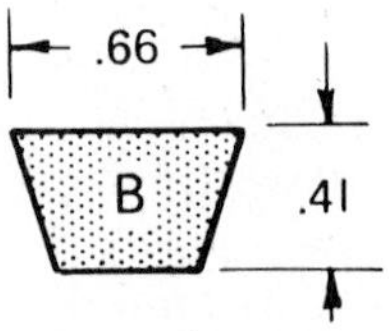

(a)

Installation allowance	take-up	belt length	sum of both v-belt pulley diameters																							
			4	4½	5	5½	6	6½	7	7½	8	8½	9	9½	10	10½	11	11½	12	12½	13	13½	14	14½	15	15½	16
½	½	16	4.9	4.5	4.1																						
⅝	½	18	5.9	5.5	5.1	4.6																					
⅝	½	20	6.9	6.5	6.1	5.6	5.2																				
⅝	½	22	7.9	7.5	7.1	6.6	6.2	5.8																			
⅝	½	24	8.9	8.5	8.1	7.6	7.2	6.8	6.3	5.8																	
⅝	½	26	9.9	9.5	9.1	8.6	8.2	7.8	7.3	6.9	6.5																
⅝	½	28	10.9	10.5	10.1	9.6	9.2	8.8	8.4	7.9	7.6	7.1	6.6														
⅝	½	30	11.9	11.5	11.1	10.6	10.2	9.8	9.4	8.9	8.6	8.1	7.7	7.3													
⅝	½	32	12.9	12.5	12.1	11.6	11.2	10.8	10.4	10.0	9.6	9.1	8.7	8.4	8.0												
⅝	½	34	13.9	13.5	13.1	12.7	12.2	11.8	11.4	11.0	10.6	10.2	9.7	9.4	9.0	8.6											
⅝	½	36	14.9	14.5	14.1	13.7	13.2	12.8	12.4	12.0	11.6	11.2	10.7	10.4	10.0	9.6	9.0										
⅝	½	38	15.9	15.5	15.1	14.7	14.2	13.8	13.4	13.0	12.6	12.2	11.8	11.4	11.0	10.6	10.0	9.7	9.1								
¾	½	40	16.9	16.5	16.1	15.7	15.3	14.8	14.4	14.0	13.6	13.2	12.8	12.4	12.0	11.6	11.1	10.7	10.1	9.8							
¾	½	42	17.9	17.5	17.1	16.7	16.3	15.8	15.4	15.0	14.6	14.1	13.8	13.4	13.1	12.6	12.1	11.7	11.2	10.8	10.2						
¾	½	44	18.9	18.5	18.1	17.7	17.3	16.8	16.4	16.0	15.6	15.2	14.8	14.4	14.1	13.6	13.1	12.8	12.2	11.9	11.2	10.9					
¾	½	46	19.9	19.5	19.1	18.7	18.3	17.9	17.4	17.0	16.6	16.2	15.8	15.4	15.1	14.6	14.1	13.8	13.2	12.9	12.3	12.0	11.3	10.9	10.5		
¾	½	48	20.9	20.5	20.1	19.7	19.3	18.9	18.4	18.0	17.7	17.2	16.8	16.4	16.1	15.6	15.1	14.8	14.3	13.9	13.3	13.0	12.3	12.0	11.6	11.3	
¾	½	50	21.9	21.5	21.1	20.7	20.3	19.9	19.4	19.0	18.7	18.2	17.8	17.4	17.1	16.7	16.2	15.8	15.3	14.9	14.4	14.0	13.3	13.1	12.7	12.4	12.1
¾	½	52	22.9	22.5	22.1	21.7	21.3	20.9	20.4	20.0	19.7	19.2	18.8	18.4	18.1	17.7	17.2	16.8	16.3	15.9	15.4	15.0	14.3	14.1	13.8	13.5	13.1
¾	½	54	23.9	23.5	23.1	22.7	22.3	21.9	21.4	21.0	20.7	20.2	19.8	19.4	19.1	18.7	18.2	17.8	17.3	17.0	16.4	16.1	15.2	14.8	14.5	14.2	13.8
¾	½	56	24.9	24.5	24.1	23.7	23.3	22.9	22.4	22.0	21.7	21.2	20.8	20.4	20.1	19.7	19.2	18.8	18.3	18.0	17.4	17.1	16.2	15.9	15.6	15.2	14.9
¾	½	58	25.9	25.5	25.1	24.7	24.3	23.9	23.4	23.0	22.7	22.2	21.8	21.4	21.1	20.7	20.2	19.8	19.3	19.0	18.5	18.1	17.3	16.9	16.6	16.3	15.9
⅞	¾	60	26.9	26.5	26.1	25.7	25.3	24.9	24.5	24.0	23.7	23.2	22.8	22.4	22.1	21.7	21.2	20.8	20.4	20.0	19.5	19.1	18.3	18.0	17.6	17.3	17.0
⅞	¾	62	27.9	27.5	27.1	26.7	26.3	25.9	25.5	25.0	24.7	24.3	23.8	23.4	23.1	22.7	22.2	21.8	21.4	21.0	20.5	20.1	19.4	19.0	18.7	18.3	18.0
⅞	¾	64	28.9	28.5	28.1	27.7	27.3	26.9	26.5	26.0	25.7	25.3	24.8	24.4	24.1	23.7	23.2	22.9	22.4	22.0	21.5	21.1	20.4	20.0	19.7	19.4	19.1
⅞	¾	66	29.9	29.5	29.1	28.7	28.3	27.9	27.5	27.0	26.7	26.3	25.9	25.4	25.1	24.7	24.2	23.9	23.4	23.0	22.5	22.2	21.4	21.1	20.7	20.4	20.0
⅞	¾	68	30.9	30.5	30.1	29.7	29.3	28.9	28.5	28.1	27.7	27.3	26.9	26.4	26.1	25.7	25.2	24.9	24.4	24.0	23.5	23.2	22.4	22.1	21.7	21.4	21.0
⅞	¾	70	31.9	31.5	31.1	30.7	30.3	29.9	29.5	29.1	28.7	28.3	27.9	27.4	27.1	26.7	26.2	25.9	25.4	25.0	24.5	24.2	23.5	23.1	22.8	22.3	22.1
⅞	¾	72	32.9	32.5	32.1	31.7	31.3	30.9	30.5	30.1	29.7	29.3	28.9	28.4	28.1	27.7	27.2	26.9	26.4	26.0	25.5	25.2	24.5	24.1	23.8	23.4	23.1
⅞	¾	74	33.9	33.5	33.1	32.7	32.3	31.9	31.5	31.1	30.7	30.3	29.9	29.4	29.1	28.7	28.2	27.9	27.4	27.0	26.5	26.2	25.5	25.1	24.8	24.4	24.1
⅞	¾	76	34.9	34.5	34.1	33.7	33.3	32.9	32.5	32.1	31.7	31.3	30.9	30.4	30.1	29.7	29.2	28.9	28.4	28.0	27.6	27.2	26.5	26.2	25.8	25.5	25.1
⅞	¾	78	35.9	35.5	35.1	34.7	34.2	33.9	33.5	33.1	32.7	32.3	31.9	31.4	31.1	30.7	30.2	29.9	29.4	29.0	28.6	28.2	27.5	27.2	26.8	26.5	26.1
⅞	¾	80	36.9	36.5	36.1	35.7	35.3	34.9	34.5	34.1	33.7	33.3	32.9	32.4	32.1	31.7	31.3	30.9	30.4	30.0	29.6	29.2	28.6	28.2	27.9	27.5	27.1
⅞	¾	82	37.6	37.1	36.7	36.3	35.9	35.5	35.1	34.7	34.3	33.9	33.5	33.0	32.7	32.3	31.9	31.5	31.0	30.7	30.2	29.8	29.2	28.8	28.5	28.1	27.8
⅞	¾	84	38.9	38.5	38.1	37.7	37.3	36.9	36.5	36.1	35.7	35.3	34.9	34.4	34.1	33.7	33.3	32.9	32.4	32.1	31.6	31.2	30.6	30.2	29.9	29.5	29.2

(b)

Fig. 3-3. (a) V-belt horsepower ratings (normal duty); (b) V-belt length.

SOLUTION

Step 1
Select driver V-pulley and belt cross section from Fig. 3-3(a). Read down the extreme left column to the rpm figure nearest to that of the speed of the motor, which is 1750 rpm. Read across this line to a number just over the design horsepower of the drive. The closest horsepower rating is .51. Read up from the hp figure. The figure at the top of the column is the pitch diameter of the motor pulley in inches. The .51-hp figure is in the white area. The reference at the bottom of the chart indicates the size of the belt required. Therefore

$$\text{Pulley size for motor} = \phi 2.75 \text{ in.}$$

Belt section = .50 in. wide × .31 in. thick, Type A

Step 2
Next, select the large pulley. Do this by inversely proportioning the speed change to the pulley diameters.

$$\text{rpm}_{in}/\text{rpm}_{out} = d_{out}/d_{in} \qquad 3\text{-}1$$

$$1750/1200 = d_{out}/2.75$$

$$d_{out} = 4.01 \text{ in.}$$

use a $\phi 4.00$-in. large pulley

Since the pulley size had to be adjusted to the closest standard diameter, the actual output speed must be recalculated. Resolving the above proportion

$$\text{rpm}_{in}/\text{rpm}_{out} = d_{out}/d_{in}$$

$$1750/\text{rpm}_{out} = 4.00/2.75$$

$$\text{rpm}_{out} = 1203$$

Step 3
Determine the belt length and center distance from Fig. 3-3(b). Add the diameter of the pulleys (2.75 + 4.00 = 6.75). Select the number in the top row that is nearest to this sum. Interpolating the table between 6.50 and 7.00 indicates a belt length of 52 in. with an approximate center distance of 20.65 in. Thus, the drive system is:

A size × 52-in.-long V-belt, $\phi 2.75$-in. V-pulley, and $\phi 4.00$-in. V-pulley.

Standard-size pulleys may be selected from the tables in Appendix 2. Using the closest A size pitch diameter, the part numbers selected are BK34 ($\phi 2.80$) and BK50 ($\phi 4.00$).

If a length of belt exceeds the values shown in Fig. 3-3(b), then the following formula must be used:

$$l = 1.57 \, (d_1 + d_2) + 2C + (d_2 - d_1)^2 / 4C \qquad 3\text{-}2$$

where l = belt length (in.)

$\quad$ d_1 = small pulley diameter (in.)

$\quad$ d_2 = large pulley diameter (in.)

$\quad$ C = center distance (in.)

From the result, select the closest available standard belt from Appendix 2 (B 50).

Multiple Strand

SPECIFICATION
Design a multiple V-belt system to transmit 2.00 hp from 1750 rpm to 1200 rpm. The shafts have an approximate center distance of 20 in.

Step 1
The horsepower rating exceeds the single-strand ratings listed in Fig. 3-3. Thus a multiple-strand system as shown in Fig. 3-4 must be used. Several possibilities exist including:

Four type A belts and $\phi 2.75$ pulleys—allow hp = 2.04 (.51 × 4)

Three type A belts and $\phi 3.25$ pulleys—allow hp = 2.22 (.74 × 3)

Two type B belts and $\phi 4.00$ pulleys—allow hp = 2.16 (1.08 × 2)

Select the one closest to the design horsepower:

Four type A and $\phi 2.75$ pulleys

Step 2
The remainder of the solution will be the same as in the section "Single Strand." Thus the drive system is:

Four A size × 52-in.-long V-belts

$\phi 2.75$-in. V-pulleys

$\phi 4.00$-in. V-pulleys

Fig.3-4. Multiple V-belt.

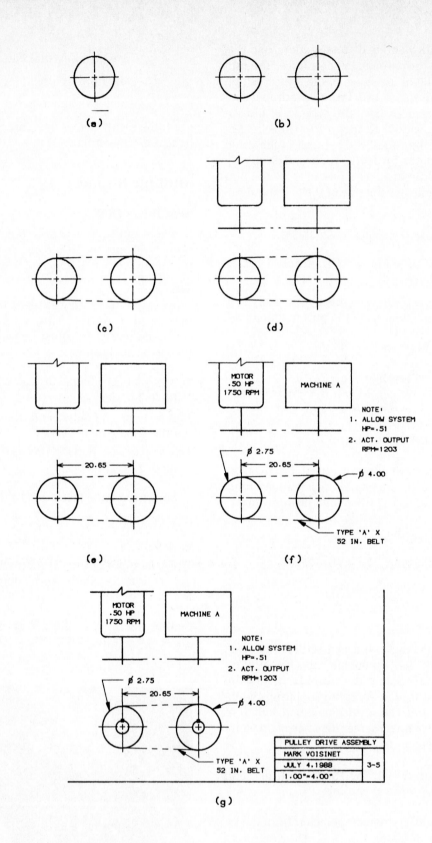

Fig.3-5 (a–g). V-belt pulley drive.

CADD DRAWING

The drive components designed in the section "Single Strand" may be shown graphically by either a double line or schematic representation. The procedure to use for a schematic presentation is:

1. Activate the CADD system. Specify a B size drawing and half scale.
2. Determine the center location of the driver (smaller) pulley. Remember to allow enough room for a top (plan) view. Draw the ϕ2.75-in.-pitch diameter using the CIRCLE command.
3. Select the ARC CENTERLINE command. Pick the arc. The centerline will automatically be placed as shown in Fig. 3-5(a). If this command is not available on your system, create each centerline individually.
4. Determine the center location for the driven (larger) pulley. It is 20.65 in. from the smaller pulley center. Remember your drawing is half scale. Use the RELATIVE or POLAR COORDINATE command to locate it.
5. Draw the ϕ4.00-in. larger pulley using the CIRCLE command. Create the centerlines with the ARC CENTERLINE or CENTER linetype command. The result will be as shown in Fig. 3-5(b).
6. Select the PHANTOM LINES and TANGENT ARCS commands. Select toward the upper part of each arc. Repeat the selection at the lower part of each arc. The result will be as shown in Fig. 3-5(c).

Note: If a double-line scale drawing is to be drawn, create the .31-in. belt thickness with a LINE OFFSET command. The lengths may be modified to each circle circumference with an EDIT LINE LENGTH command.

7. Create the top view. Use a combination of CONSTRUCTION LINES, LINES, FILLETS, and PHANTOM commands. Estimate the motor and machine sizes. The result will be as shown in Fig. 3-5(d).
8. Dimension the center to center distance with HORIZ DIM. The result is as shown in Fig. 3-5(e).
9. Create the remainder of the drawing using RADIAL DIM and TEXT. The result is as shown in Fig. 3-5(f).

10. If the pulley bore and keyway sizes are known, these may be added to the drawing.
11. Finish the assembly by adding a title block. If a title block is not a standard symbol on your system, the example shown in Appendix 6 may be used. The result is shown in Fig. 3-5(g).

ASSIGNMENTS

1. A 0.75-hp, 1750-rpm motor drives a machine whose flywheel turns at approximately 500 rpm and operates under normal conditions. A pulley drive system is used to transmit the power and motion between the motor and flywheel shaft.
   a. Determine the required drive system.
   b. Create a single-line schematic assembly diagram specifying each drive component, allowable horsepower, and actual output rpm.
2. Repeat assignment 1 for a 1.25-hp, 1750-rpm motor. The center-to-center shaft distance is approximately 13.50 in., and the output rpm is approximately 800.
3. Repeat assignment 1 for a 1.00-hp, 1160-rpm motor. The center-to-center shaft distance is the minimum, and the output rpm is approximately 415.
4. Determine the allowable horsepower ratings of the pulley drives shown in Fig. 3-6.
5. Determine the allowable horsepower ratings of the pulley drives shown in Fig. 3-7.
6. A belt drive system transmits 1.00 hp at 1160 rpm. The output shaft is located approximately 24 in. from the motor shaft and rotates approximately 2.25 times slower than the input shaft. Select the appropriate V-belt and drive pulley system. Use Figs. 3-3 (a) and (b) to design the system operating under normal duty. Prepare an assembly drawing of the drive system. Include ϕ1.00-in. bores and keyways for .25-in. square keys in each pulley.
7. The 7.50-hp, 1750-rpm motor shown in Fig. 3-8 drives a centrifugal blower. The fan has a 12-in. diameter and is approximately 6.00 in. wide. It is to turn at as close to 550 rpm as possible. This system operates under normal duty.
   a. Design the pulley and belt drive. Use a multiple-belt system.
   b. Prepare an assembly drawing of the drive system.

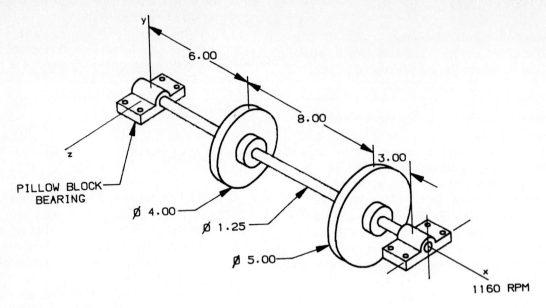

Fig.3-6. V-belt problem.

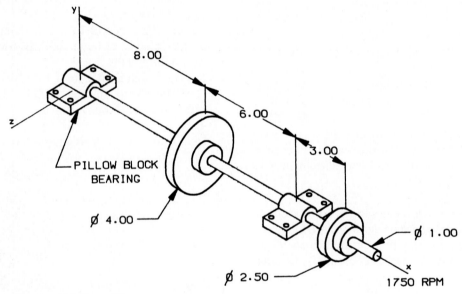

Fig.3-7. V-belt problem.

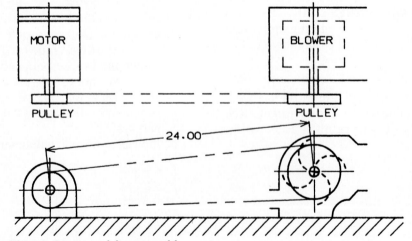

Fig.3-8. Motor and fan assembly.

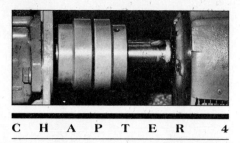

SHAFT/KEY/ COUPLING

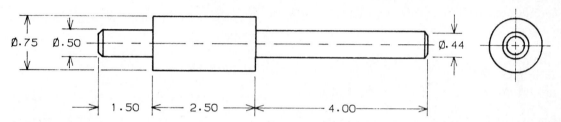

Fig. 4-1. *Shaft.*

DESIGN THEORY

Shafts

A shaft is a mechanical component normally having a circular cross section (Fig. 4-1). It is used in conjunction with other mechanical power transmitting components such as pulleys, gears, sprockets, and sheaves. These components have a bore diameter matching the shaft diameter to slide on the shaft for

easy assembly. Both will then rotate together transmitting motion and power.

How is the size of a shaft determined? First, shafts are subjected to various types of forces, or loads. These include tension, compression, bending, and torsion. The effects of the combination of these stresses must be taken into account. For mechanical power transmission, however, the major type of loading is torsion (Fig. 4-2). The theoretical shaft torsion formula is

$$S_s = TRK/J \qquad \text{4-1}$$

S_s = shearing stress, pounds per square inch (psi)

T = torque, pound-inches (lb-in.)

R = radius of shaft (in.)

J = polar moment of inertia (in.^{4})

K = stress concentration factor (no units)

The shearing stress (S_s) at which a shaft will fail in torsion depends on the type of material plus other variables. These variables include exact loading, shocks, corrosion, and temperature cycling. It is not usually possible to identify the magnitude of the vari-

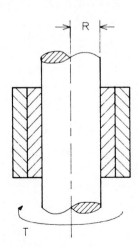

Fig. 4-2. *Torsion.*

ables. To take each unknown value into account, designers will work with a factor of safety. The amount over the safety factor will vary depending on the application. It generally, however, will be in the range of several times (for example, 3 to 4) the theoretical value.

The stress concentration factor (K) term in the above equation depends on shaft material variation. Shafts will possess some form of nonuniformity or flaws. They may have holes, keyways, turned-down sections, retaining ring grooves, or perhaps just marks. All variations will weaken the shaft to some extent. For example, a simple keyway may decrease shaft strength by as much as 60 percent. Consequently, a stress concentration factor must be taken into account.

The remaining terms in the above formula may be readily determined. The torque (T), for example, is a function of the speed (rpm) and power (hp) to be transmitted. These are usually known values. The polar moment of inertia (J) may be found in strength of material tables. (It is $\Pi D^4/32$ for a solid circular shaft.)

After the above values have been determined, the torsion formula may be solved for the theoretical required minimum radius (r) of the shaft. The result will be somewhat inaccurate for the reasons described above. Thus a more practical formula is desirable.

The following is a conservative, easy to use shaft formula:

$$D = \sqrt[3]{(80 \times hp)/n} \qquad \text{4-2(a)}$$

where D = minimum required shaft diameter (in.)

hp = system horsepower

n = rpm of that shaft

80 = constant for standard carbon steel

It takes each of the above-mentioned variables into consideration, neglects excessive deflection (bending stress), and provides a generous factor of safety. Common cold-drawn steel shafting material is used. This equation is to be applied when determining main power transmission shaft diameters. For lineshafts that carry pulleys, the required diameter may be reduced by using the formula

$$D = \sqrt[3]{(53.5 \times hp)/n} \qquad \text{4-2(b)}$$

where 53.5 is the constant for lineshafts carrying pulleys. Occasionally you may want to use the smallest possible diameter shaft. If the shaft is short, excessive twisting will not be a problem. Thus use the following formula:

$$D = \sqrt[3]{(38 \; hp)/n} \qquad \text{4-2(c)}$$

where 38 is the constant for short shafts.

Fig. 4-3(a). *Key application.*

Keys

A *key* is a small piece of metal fitting between the shaft and mechanical power-transmitting component. Its purpose is to transmit system torque between the two parts (from shaft to component or component to shaft) without failure as shown in Fig. 4-3(a). It also prevents slipping or spinning. Thus keys provide positive transfer of power and motion between shaft and moving part(s).

When selecting a key, two factors must be determined. These are:

■ Size
■ Type

Usual practice is to choose standard-size key stock, which is approximately one-fourth of the shaft diameter. For example, a φ1.00-in. shaft will utilize a .25-in. square key (maximum). Refer to the first table in Appendix 3 for other sizes. Next, the key length must be determined. This becomes a function of the power to be transmitted. For low-horsepower and/or high-rpm applications the length is not a critical factor. Just be certain that it is not too short (usually at least .50 in.) because assembly can be difficult. Also, it should not exceed the length of the machine element. For high-torque (high-horsepower, low-rpm) applications, however, the minimum length must be calculated.

There are several types of keys available. The common ones illustrated in Fig. 4-3(b) and specified in Appendix 3 include:

■ Square
■ Flat
■ Gib-head
■ Pratt and Whitney
■ Woodruff

TYPE OF KEY	ASSEMBLY SHOWING KEY, SHAFT AND HUB	SPECIFICATION
SQUARE		.25 SQUARE KEY, 1.25 LG .25 SQUARE TAPERED KEY, 1.25 LG
FLAT		.188 X .125 FLAT KEY, 1.00 LG .188 X .125 FLAT TAPERED KEY, 1.00 LG
GIB-HEAD		.375 SQUARE GIB-HEAD KEY, 2.00 LG
PRATT AND WHITNEY		NO. 15 PRATT AND WHITNEY KEY
WOODRUFF		NO. 1210 WOODRUFF KEY

Fig.4-3 (b). Common keys.

Selection depends upon the application. Square keys are used with keyseats at the end of shafts. Flat when a small radial distance (such as a small hub diameter) is required. Some situations require key placement near the center of a long shaft. It is not feasible to cut a key slot all the way from one end. Thus a woodruff key would be used as shown in Fig. 4-3(b).

After the type of key has been selected, determine the torque to be transmitted (Eq. 4-3 below). If this is significant, good design practice dictates that the minimum key length procedure be followed. The equations and assumptions for key selection are as follows.

Step 1
If the horsepower and the rotational speed of the shaft are known, torque can be determined from:

$$T = (63{,}025 \times hp)/n \qquad \text{4-3}$$

where T = torque (lb-in.)

hp = system horsepower

n = rpm of that shaft

63,025 = constant

Step 2

The force acting on the key can be determined from:

$$P = T/R \qquad \text{4-4}$$

where P = actual force (lb)

T = torque (lb-in.)

R = radius of shaft (in.)

This force is between the two mating parts. It is assumed to be uniformly distributed along the length of the key as illustrated in Fig. 4-4.

The required length of the key can now be determined. The shear area (A) and the allowable shear stress (S) need to be defined. The shear area is the key width (t) times the length (l) at the line of contact between the shaft and the machine element. This is illustrated in Fig. 4-4(a) and is equal to t × l.

Step 3
The final computation used to determine minimum key length is:

$$S = P/A = P/tl \qquad \text{4-5}$$

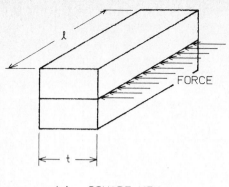

(a) SQUARE KEY

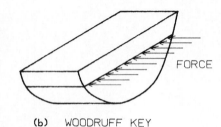

(b) WOODRUFF KEY

Fig. 4-4 (a and b). Shear area.

Therefore the minimum key length is

$$l = P/tS$$

where S = allowable shear stress (psi) = 8,000 psi

P = transmitted force (lb)

A = shear area (in.^{2})

t = key thickness (in.)

l = minimum required key length (in.)

A reasonable allowable shear stress of 8000 psi is used. This will provide for a factor of safety of approximately 4 if standard steel stock is used.

This selection method works equally well for square, flat, gib, Pratt and Whitney, and woodruff keys. Even though the shear area of a woodruff key (Fig. 4-4[b]) is less, it is not significant. As previously stated, the minimum required length becomes a critical factor for low-rpm and high-horsepower applications. In fact, if the minimum required key length exceeds the machine component hub length of two keys, another method such as splining may be required.

Couplings

Often, more than one shaft is used for a specific application. It may be necessary to join two different ones. Couplings are used for this purpose. They provide the option to connect shafts of equal or different (but close) diameters. Standard couplings may be selected from a chart (or manufacturer's catalog) as illustrated in Appendix 4. Select the coupling which will fit the range of both shaft diameters. Be sure to check the maximum horsepower rating in the chart. Compare the maximum allowable torque value to required torque (Eq. 4-3). The coupling selected will only be able to safely withstand the specified torque.

DESIGN APPLICATION

Shaft Selection

SPECIFICATION

Select a shaft capable of transmitting 7.50 hp rotating at 100 rpm.

SOLUTION

Minimum diameter using Eq. 4-2(a):

$$D = \sqrt[3]{(80) \times (7.50)/100} = \sqrt[3]{6.000} = \phi 1.816 \text{ in.}$$

Since a ϕ 1.816-in. is not a standard stock size (refer to Appendix 5) the next higher diameter should be selected. Thus use ϕ1.88 (or 2.00) in. If the shaft is not the main power shaft but a lineshaft carrying pulleys, use Eq. 4-2(b). When solving Eq. 4-2(b), we find that a ϕ1.62 in. may be used.

Key Selection

SPECIFICATION

Determine the required size for a square key. It must be capable of transmitting 7.50 hp at 100 rpm between a ϕ2.00 $\times$ 8.00 in. (e.g., shaft and a pulley).

SOLUTION

For the torque

$$T = (63,025) \times (7.50)/100 = 4726 \text{ in.-lb}$$

For the force

$$P = 4726/1.00 = 4726 \text{ lb}$$

For the minimum length and for a 2.00-in. shaft, a .50-in. square key may be used (refer to Appendix 5 for standard sizes). Using an allowable shear stress of 8000 psi

$$l = 4726/(8000) \times (.50) = 1.18\text{-in. minimum length}$$

Use a .50-in. square $\times$ 1.25 long key.

Coupling Selection

SPECIFICATION

Select a coupling to join ϕ1.88-in. and ϕ2.00-in. shafts. The coupling must be capable of transmitting 7.50 hp at 100 rpm.

SOLUTION

For the torque

$$T = (63{,}025) \times (7.50)/100 = 4726 \text{ in.-lb}$$

Refer to Appendix 4. The chart shows that shaft diameters in the range of 1.75 to 2.12 have a maximum allowable torque of 11,000 in.-lb. (17.50 hp at 100 rpm). This is well over the torque developed by the system.

SPECIFICATION

Type FC45—1.88 × 2.00 in.

CADD DRAWING

Create a working drawing of the shafts, key, and coupling designed in the sections "Shaft Selection,"
"Key Selection," and "Coupling Selection." Prepare the drawing as follows:

1. Create the drawing file and establish the units (B size and full scale).
2. Select the CENTERLINE (or LINE and CENTER layer) command.
3. Create a horizontal centerline 14.00 in. long as shown in Fig. 4-5(a).
4. Gather the coupling size data from Appendix 4.
5. Create the outline of the coupling using LINES (RELATIVE COORDINATE) and FILLET commands. Begin drawing at an X coordinate of 3.00 in. as shown in Fig. 4-5(a).
6. Create the ϕ1.88-in. shaft extending from the left of the coupling. Since the length is unknown, show a conventional break. Use the LINES, (S) PLINE, and HATCH commands. The result will be as shown in Fig. 4-5(b).

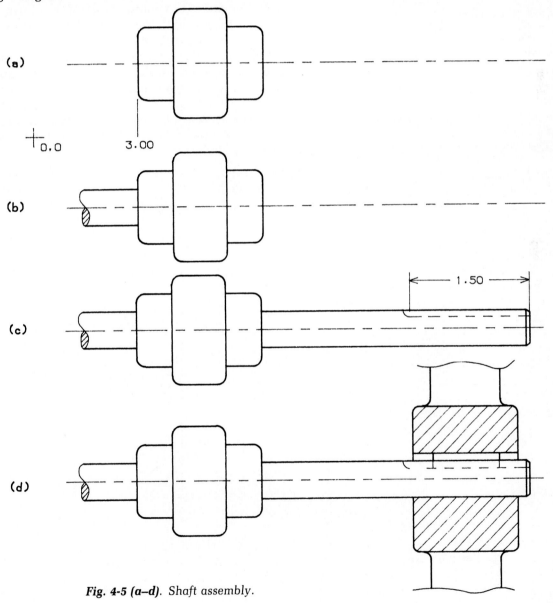

Fig. 4-5 (a–d). Shaft assembly.

7. Create the φ2.00- × 8.00-in.-long shaft using the LINE command. Subtract the coupling hole length to determine the extension. Include a .12 × 45° chamfer (use a CHAMFERS command if available).

8. Use HIDDEN and FILLET commands to create a 1.50-in.-long sled runner keyway as shown in Fig. 4-5(c). Remember the depth of a keyseat for a square key is normally t/2.

9. Complete the drawing by adding the 1.25-in.-long key and a mechanical component hub of φ5.00 × 3.00 in. long. Use the sequence of LINES, FILLETS, BLOCK, and HATCH commands for each hub half. The result is shown in Fig. 4-5(d).

ASSIGNMENTS

1. Create an assembly drawing of a φ.50-in. shaft, two .12 square keys and mechanical components as shown in Fig. 4-6.

2. a. Design two 6.00-in.-long shafts capable of transmitting 5 hp at 400 rpm.
   b. Determine the square key required to transmit the system torque.
   c. Select a coupling capable of joining the two shafts.
   d. Create an assembly drawing of the shafts, coupling, and key. Place the key at the end of one shaft. Add .12 × 45° chamfers on each shaft.

3. a. Select a coupling capable of joining φ1.25- and φ1.50- × 5.00-in.-long shafts.
   b. Determine the allowable horsepower of the smaller shaft if it rotates at 585 rpm.
   c. Create an assembly drawing of the shafts and coupling. Select an appropriate key of minimum length for each shaft. Include both on the drawing.

4. Shaft/Coupling Problems
   Determine the allowable horsepower for the shafts shown in Fig. 4-7(a) and (b). Break each into two shafts at the step-down location. Select the appropriate coupling to rejoin them. Create a sectional assembly drawing of the shafts and couplings as illustrated in Fig. 4-7(c) and (d).

5. Continuation of assignment 3, Chapter 3.
   a. Determine the required size of the fan shaft.
   b. Determine the required size of the square key to be used with the pulley on the fan shaft.
   c. Prepare an assembly drawing of the fan, fan shaft, pulleys, belts, key, motor, and housing.
   d. Add notes to the drawing indicating the results of each design calculation.
   e. Prepare an items list for each part except the fan. This will be selected by others. Use the title block and item list format as illustrated in Appendix 6.

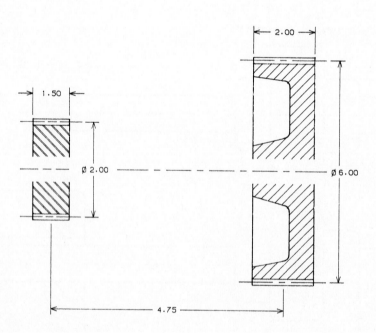

Fig. 4-6. Shaft/key problem.

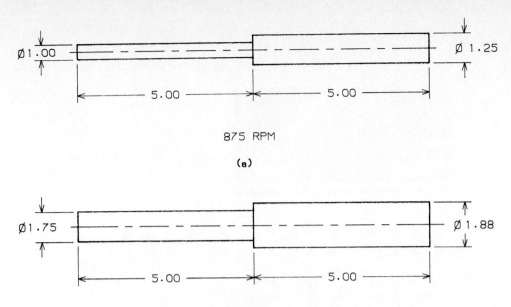

Ø1.00   Ø 1.25

5.00   5.00

875 RPM

(a)

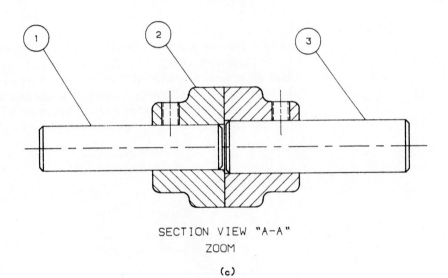

Ø1.75   Ø 1.88

5.00   5.00

600 RPM

(b)

1   2   3

SECTION VIEW "A-A"
ZOOM

(c)

(d)

Fig. 4-7 (a–d). Shaft/key problems.

SHAFT / KEY / COUPLING

C H A P T E R   5

BEARINGS AND SEALS

DESIGN THEORY

Bearings are required on virtually every power transmitting machine. They are assembled in a housing and used to support rotating shafts. Common bearing types include:

- Journal
- Ball
- Roller
- Thrust
- Needle

A typical ball bearing assembly is shown in Fig. 5-1. Bearings are shown on an assembly drawing in a variety of ways. Common representations are illustrated in Fig. 5-2.

After the fundamentals of bearings and statics are understood, one can consider selection criteria. The design and sizing of bearings depends upon shaft diameter, loads to which bearings are subjected, and speed (rpm). Primary loading is normally an axial force.

Typical bearing mountings are shown in Fig. 5-3(a). If bearings are encased in an oil-filled housing, seals will also be required. A typical application of a bearing and seal application is shown in Fig. 5-3(b).

DESIGN APPLICATION

Before proceeding, review the fundamentals of bearings and moments from any traditional text.

EXAMPLE 1

Step 1
Consider the drive component transmitting its total force to the mating component at a concentrated load. If this component is located between two bearings, take summation of moments to determine the resultant axial force. These axial forces keep the system in equilibrium. Locating the component centrally yields

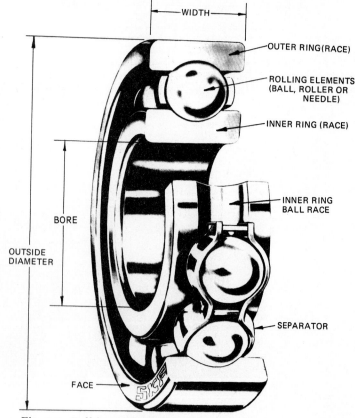

Fig. 5-1. Ball bearing (SKF Co.).

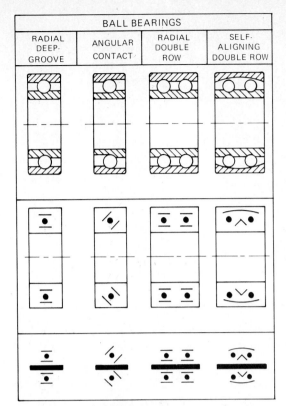

Fig. 5-2. Representation of bearings.

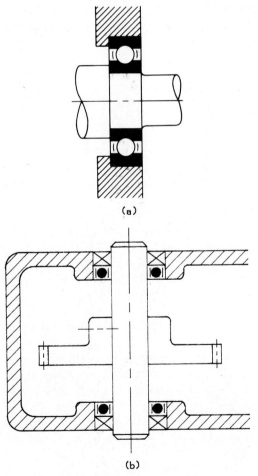

(a)

(b)

Fig. 5-3 (a and b). Casting assembly.

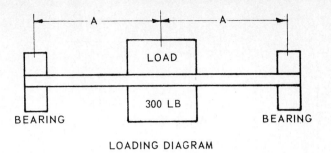

LOADING DIAGRAM

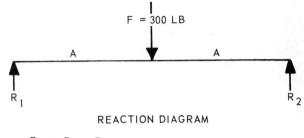

REACTION DIAGRAM

$$F_1 = R_1 + R_2$$

$$R_1 = \frac{300}{2} = 150 \text{ LB}$$

Fig. 5-4. Load located centrally on bearings.

forces on each bearing equal to half the component force and opposite in direction as shown in Fig. 5-4.

Step 2

If the component is not located between the bearings, a cantilever effect exists and resulting bearing loads will be unbalanced. The axial forces likewise will be determined by summation of moments as shown in Fig. 5-5. Note that for the same transmitted force, much larger bearings are required for this condition.

Step 3

After the axial forces to which the bearings are subjected have been determined, and with the shaft speed known, bearings may be selected. Use tables from manufacturer catalogs similar to those of Appendix 7. For the bearing loads determined by Step 1 at a shaft diameter of .50 in. rotating at 1200 rpm, bearing No. NR1616 may be selected. These bearings will withstand a radial load of 185 lb at 1200 rpm. The average life of the bearings at that load is 2500 h. Average life means that 50 percent of the bearings will exceed the stated life value.

Step 4

If a longer bearing life is desired, select one with a greater radial capacity. Life varies according to:

$$N = \text{average life } (C/Pe)^3$$

where N = rated life (hours)

C = rated bearing load (lb)

Pe = equivalent actual radial load (lb)

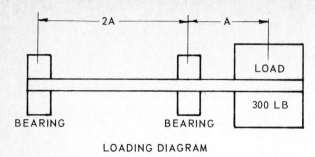

LOADING DIAGRAM

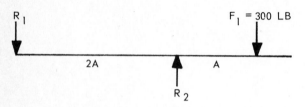

REACTION DIAGRAM

$$R_2 = F_1 + R_1$$

$$R_1 \times 2A = F_1 \times A$$

$$R_1 = \frac{F_1 \times A}{2A} = \frac{300 \times A}{2A} = 150 \text{ LB}$$

$$R_2 = F_1 + R_1 = 150 + 300 = 450 \text{ LB}$$

Fig. 5-5. *Load located outside of supporting bearings.*

For example, if bearing No. 1621 is selected, the life will be extended.

$$N = 2500 \ (300/150)^3 = 20,000 \text{ hours (average)}$$

Since average life varies with the cube of the load quotient, it is significantly increased by small increases in the rated load.

Some manufacturers specify bearing life by B-10 rating. This rating means longer bearing life and higher reliability, since 90 percent of the bearings will exceed the life and 50 percent will exceed 5 times the life.

EXAMPLE 2

The force (pounds) from the drive component will be different depending on the component being used. For example:

Gears F_{tang} = torque/radius of pitch diameter

$F_{total} = F_{tang}/\cos$ pressure angle
(as shown in Fig. 5-6)

Chain F_{total} = torque/radius of pitch diameter

Belt F_{total} = 2 × torque/radius of pitch diameter

Belt drives exert the greatest force, since tension exists in the drive and slack sides. Approximate the total force as double.

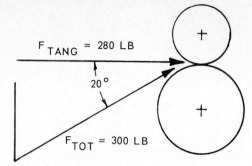

Fig. 5-6. *Transmitted forces developed from spur glass in mesh.*

Step 1

The torque exerted by a 6P, 48N, 20° spur gear is 1120 in.-lb. Determine the total component force.

$$\text{Pitch diameter} = N/P = 48/6 = 8.00 \text{ in.}$$
$$\text{and } \cos 20° = .94$$

$$F_{tang} = \text{torque/pitch radius} = 1120/4.00 = 280 \text{ lb}$$

$$F_{total} = 280/.94 = 300 \text{ lb (as used in Example 1, Step 1)}$$

For an 8.00-in. sprocket chain drive system, the total force would be 280 lb (1120/4.00).

For an 8.00-in. pulley belt drive system, the total force would be approximately 560 lb (2 × 1120/4.00).

CADD DRAWING

Create a section plan view working drawing of the bearings and shaft selected in Example 1, Step 3. Add a centrally located $\phi 3.00$ pitch diameter wide pulley #BK36 (Appendix 2). Place a housing (partial) around the assembly and include seals. Prepare the drawing as follows:

1. Boot the system, create a drawing file, and SET-UP the units (size and scale).
2. Select the GRIDS command.
3. Select and display a .25-in. grid pattern. Be sure that the grid SNAP is on.
4. Select the CENTERLINE command.
5. Draw a 4.00-in.-long vertical centerline at the center of the screen as shown in Fig. 5-7(a).
6. Next, determine the required shaft length. Add the length of each component plus .50-in. space between the pulley and casting wall. The width of an NR1616 bearing is .38 (Appendix 7), the seal width is .25 (Appendix 8), and the width of a BK36 pulley is 1.30 (Appendix 2). Thus the shaft length will be:

1.30 + (.50 × 2) + (.38 × 2) + (.25 × 2)

= 3.56 in. long minimum

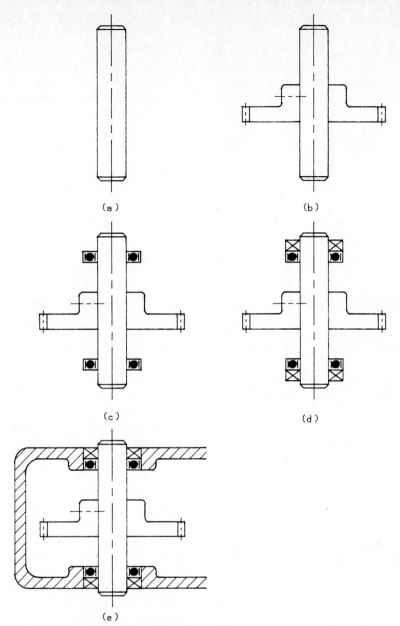

Fig. 5-7 (a–e). Bearing assembly CADD drawing.

7. Draw a φ.50- × 3.50-in.-long shaft centrally located about the centerline.

8. Create .12 × 45° chamfers on each end of the shaft. Use a CHAMFER command if available. If not, ZOOM in, change the GRID to .12 and draw using the LINE command. The result will appear as shown in Fig. 5-7(a).

9. Determine the pulley sizes from Appendix 2. Use your judgment for sizes not given. Create the outline of the pulley on the center of the shaft. Use the LINE, CENTERLINE, and BLOCK commands. The result will appear as shown in Fig. 5-7(b).

10. Determine the bearing sizes from Appendix 7. Draw a schematic representation (Fig. 5-2) of each ball bearing using the LINE, CIRCLE, and SOLID commands. Locate the bearings .50 in. away from the pulley as shown in Fig. 5-7(c).

11. Select seals (Appendix 8) the same OD as the bearings. Locate and place the seals next to the bearings. Draw a schematic representation of each as shown in Fig. 5-7(d).

12. Create the casting as shown in Fig. 5-7(e). Use the LINE, FILLET, BLOCK, and HATCH commands. Add retaining rings as necessary.

ASSIGNMENTS

1. Create an assembly drawing of the components shown in Fig. 5-8. Select standard ball bearings from Appendix 7 by matching each bore to the shaft diameters. Select appropriate seals from Appendix 8. If a 200-lb force is acting at the center of each shaft, determine bearing life.
2. Create an assembly drawing for the schematic shown in Fig. 5-9. The two horizontal shafts are φ1.00 in. and the vertical shaft is φ.75 in. Select and specify the appropriate bearings.
3. The ends of a shaft are to be supported by two ball bearings with a V-belt drive pulley held midway between. The pulley transmits 1 hp at 1160 rpm and has a pitch diameter of φ4.50 in. The bearings are held on the shaft by means of sleeves and retaining rings. Select the bearings from Appendix 7. Make your selection so that the rated radial capacity is able to withstand the force exerted by the driver pulley. Prepare a subassembly drawing using the appropriate shaft. The pulley may be purchased with a solid hub that will be bored out.
4. Select suitable fan shaft bearings for assignment 3-7. Be sure to include design rationale and the average bearing life on the assembly drawing for that project.

5. Four bearings are used to support the shafts shown in Fig. 5-10. The bearings on shaft 1 produce a cantilever effect, while those on shaft 2 produce a simple support. The pitch diameters of the mechanical drive components are φ4.00 in. and φ12.00 in., respectively. The objective of this project is to compare bearing life using different types of drive components with the same power transmitting requirements. There will be a substantial difference. Compare the different lives for belt, chain, and 20° PA gear drive applications. In each case the input shaft is driven by a 15-hp, 575-rpm motor.

   a. Determine the actual radial forces acting on the four bearings for each type of drive.
   b. Select bearings from Appendix 7 based upon the loads determined from the belt drive situation. Try to select bearings with allowable loads close to the actual forces. Disregard bore diameter requirements.
   c. Determine the life of the bearings for each of the three types of applications.
   d. Prepare a line diagram listing this information. Include the following:
      Schematic assembly drawing.
      Specify each part (use leader line or item list format).
      Present results in tabular form on the diagram.

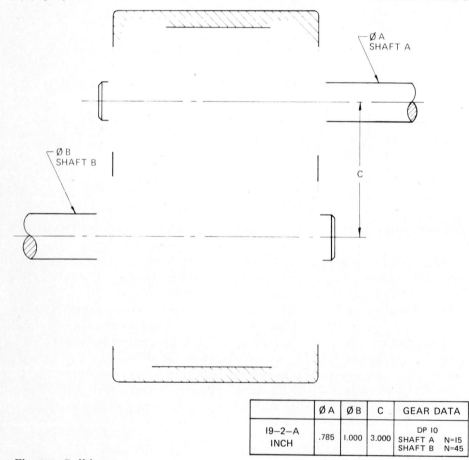

	Ø A	Ø B	C	GEAR DATA
19–2–A INCH	.785	1.000	3.000	DP 10 SHAFT A N=15 SHAFT B N=45

Fig. 5-8. Ball bearings.

MECHANICAL DESIGN USING CADD

34

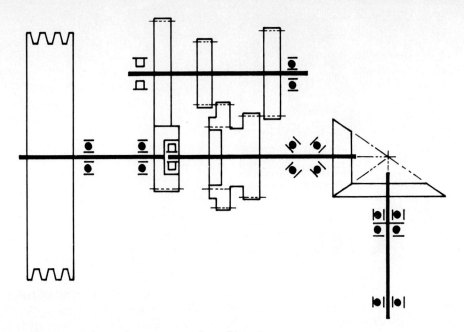

Fig. 5-9. *Schematic representation of bearings.*

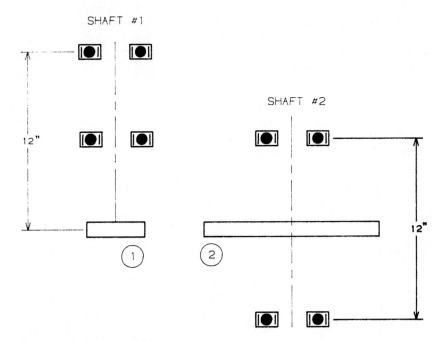

Fig. 5-10. *Bearing problems.*

GEAR

DESIGN THEORY

There are many different types of gears. The most common of these are spur gears as illustrated in Fig. 6-1. Gear design is based on strength and wear factors. Equations may be used to calculate the exact required size of a gear or gear train.

Various charts are available, however, which incorporate each variable, thus precluding the necessity to calculate tediously. A typical chart is shown in Fig. 6-2 and Appendix 9. It combines pressure an-

gle (PA), speed (rpm), pitch (P), number of teeth (N), and allowable horsepower (hp) factors.

To determine the safe capacity of a gear, first locate it on the chart by its pitch and number of teeth. Next, read the allowable horsepower rating horizontally to the left. A 20°, 12-pitch, 16-tooth spur gear turning at 900 rpm, for example, has an approximate rating of 3 hp assuming normal operating conditions. Conditions other than normal require the inclusion of a service factor (SF). This factor is determined from the Fig. 6-2 chart provided for that purpose. When

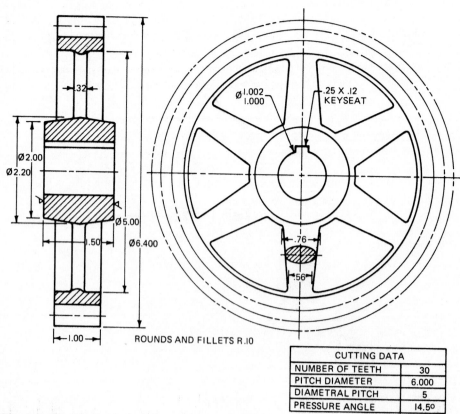

ROUNDS AND FILLETS R.10

CUTTING DATA	
NUMBER OF TEETH	30
PITCH DIAMETER	6.000
DIAMETRAL PITCH	5
PRESSURE ANGLE	14.5°

Fig. 6-1. Spur gears.

PITCH SELECTION CHART
20° PRESSURE ANGLE SPUR GEAR

*** RPM – PITCH – APPROXIMATE HORSEPOWER CHART ***
20° PA – 16 & 20 TOOTH STEEL SPUR GEARS

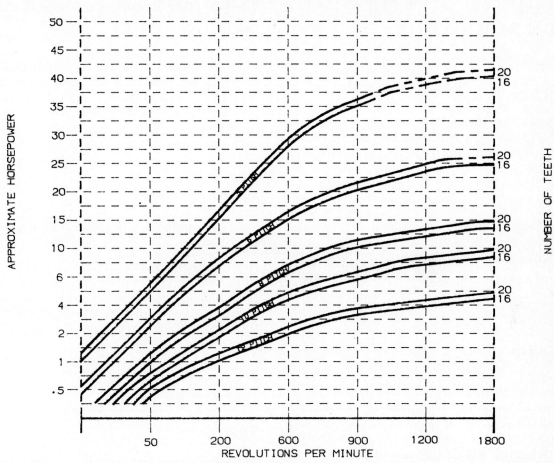

NOTE: USE ONLY THE SOLID LINE PORTION OF THE RATING CHART WHEN DETERMINING THE APPROXIMATE ALLOWABLE HORSEPOWER

(a)

BGW Spur Service Class	Operating Conditions	Service Factor [1]
Class I	Continuous 8 to 10 hrs. per day duty, with Smooth Load (No Shock).	1.0
Class II	Continuous 24 hr. duty, with Smooth Load, or 8 to 10 hrs. per day, with Moderate Shock.	1.2
Class III	Continuous 24 hr. duty, with Moderate Shock Load.	1.3
Class IV	Intermittent duty, not over 30 min. per hr., with Smooth Load (No Shock).	.7
Class V	Hand operation, Limited Duty, with Smooth Load (No Shock).	.5
Heavy Shock loads and/or severe wear conditions require the use of higher Service Factors. Such conditions may require Factors of 1.5 to 2.0 or greater than required for Class I service.		

Courtesy Boston Gear Works

SERVICE FACTOR CHART

(b)

Fig. 6-2 (a and b). *Pitch selection and service factor charts.*

determining the allowable horsepower of two gears in mesh, use the smaller (pinion) of the pair. It normally will have a lower capacity.

DESIGN APPLICATION

Spur Gears

SPECIFICATIONS

A 7.50-hp, 1200-rpm motor is used to drive a machine that runs 8 h per day under moderate shock. If the machine is to run at 200 rpm and at the capacity of the motor, what spur gears would you select? Use the chart shown in Fig. 6-2 to get an idea of the approximate size for the meshing gears.

SOLUTION

Step 1
Refer to the service factor chart (Fig. 6-2). The operating conditions are such that the machine fits into the service class No. II and requires a service factor of 1.2.

Step 2
The design horsepower becomes 7.50 × 1.2 = 9.00 hp.

Step 3
Determine the spur gear pinion rating by referring to the approximate horsepower graph in Fig. 6-2. At 1200 rpm, for a 20° pinion, the required pitch and number of teeth can be found. Reading vertically on the 1200-rpm line and horizontally at 9 hp, you can see that the 8-pitch band is located just above the intersection of these factors. The allowable horsepower of a 16-tooth pinion is well above 9.00. Referring to the table in Appendix 9, it is seen to be 10.30. Thus select a 20°, 8-P, 16-N gear.

Step 4
To determine the mating gear running at 200 rpm or 1/6 of the pinion rpm

$$\text{Number of teeth} = 16 \times 6 = 96 \text{ teeth}$$

The required pitch and pressure angle of the mating gear must remain the same. The catalog horsepower rating of the larger gear can also be approximated by using the charts in Fig. 6-2. This, however, is not necessary (except in critical cases), since the pinion normally has lower ratings.

Step 5
Select the gears from a standard manufacturer's catalog. Typical pages are shown in Appendix 9. Refer to the 8-pitch section. The catalog numbers for the gears are YH16 and YH96B.

Bevel Gears

SPECIFICATIONS

Select a pair of bevel gears capable of transmitting 1.0 hp at 600 rpm and a speed reduction of 2:1.

SOLUTION

Step 1
Refer to Appendix 10. The approximate hp ratings are listed by speed ratio and rpm. Locate the 2:1 ratio section. Refer to the 600-rpm column. Read down the column until an approximate rating of 1.0 (or just over) is found. There are two choices that fit the criteria. They are:

10 pitch (#PA5210Y)

12 pitch (#HL152Y)

Select one of the two (for example, 10 pitch).

Step 2
Next, the gear data and catalog numbers may be specified. Bevel gears are purchased as a pair. These are referred to as gear (G) and pinion (P). The associated numbers (from Appendix 10) are PA5210Y-G and PA5210Y-P. Each is 10 pitch with 50 and 25 teeth, respectively.

CADD DRAWING

Spur Gear

Create a working drawing of the gears designed in the section "Spur Gears." Refer to Appendix 9. Gather the size data given for both the 16- and 96-tooth gear. Prepare the drawing as follows:

1. Create the drawing file and SETUP the units (C size and full scale).
2. Select the GRIDS command.
3. Select and display a .25-in. grid pattern.
4. Select the CENTERLINE command.
5. Position and set the cursor at points 1 and 2 as shown in Fig. 6-3(a). Remember to MOVE NEXT POINT before selecting points 3 and 4. The centerlines will be drawn as shown.
6. Select the CIRCLE and CENTER line-type commands.
7. Position and set the cursor at the intersection of the centerlines (shown as point 5 in Fig. 6-3(a).
8. Position and set the cursor at point 6, located 24 grid dots (6.00 in.) to the right of point 1. This corresponds to the radius of the 12-in. pitch diameter of the large gear. The pitch circle will be drawn as shown.
9. Determine the outside and root diameter. Use

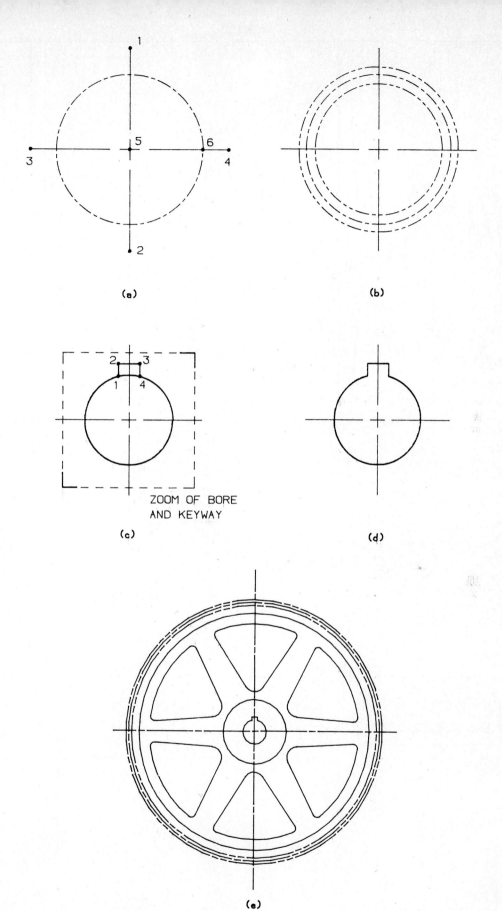

(a)

(b)

ZOOM OF BORE
AND KEYWAY

(c)

(d)

(e)

Fig. 6-3 (a–e). Spur gear working drawing.

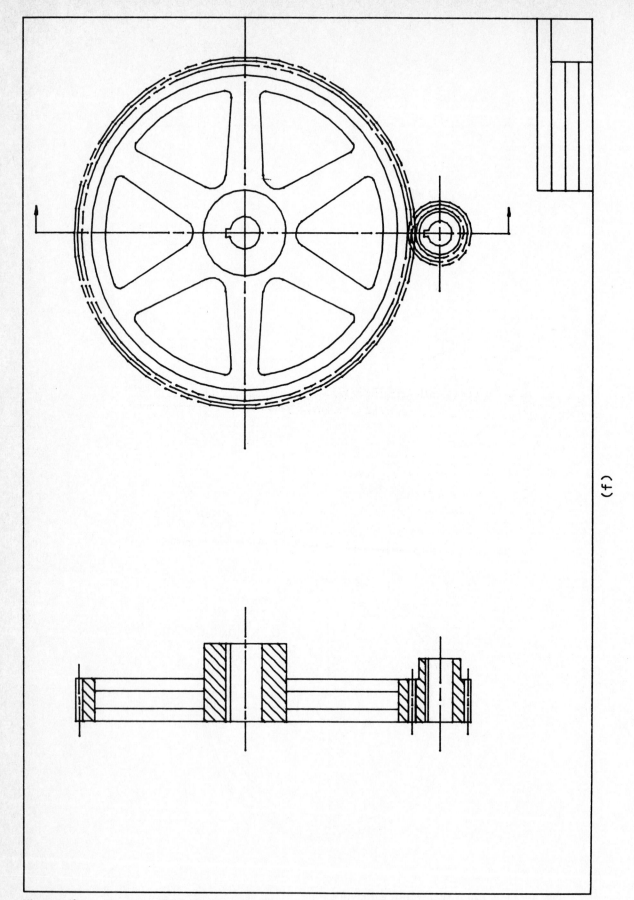

Fig. 6-3 (f). Spur gear working drawing (cont.).

(f)

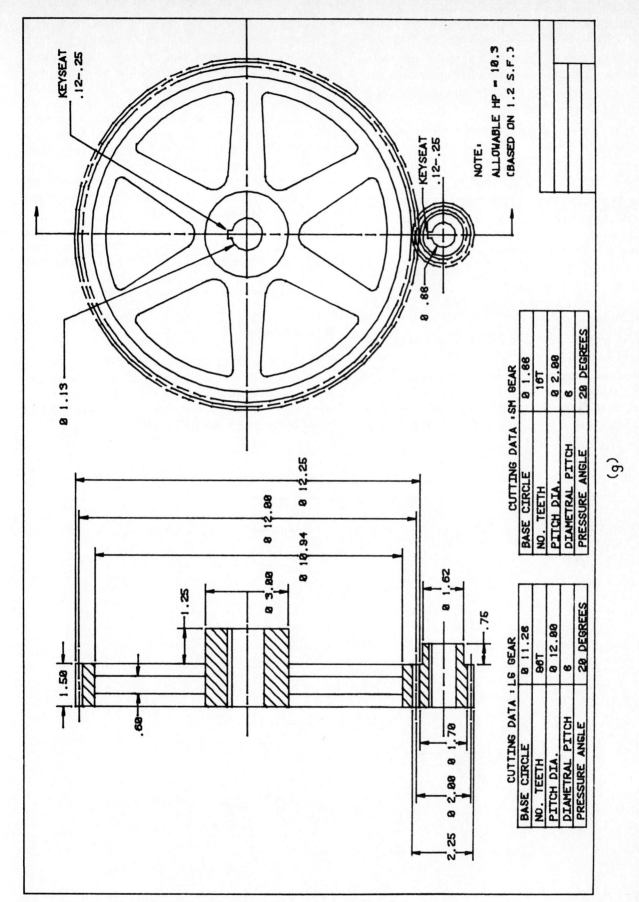

CUTTING DATA : SM GEAR

BASE CIRCLE	Ø 1.88
NO. TEETH	16T
PITCH DIA.	Ø 2.00
DIAMETRAL PITCH	8
PRESSURE ANGLE	20 DEGREES

CUTTING DATA : LG GEAR

BASE CIRCLE	Ø 11.28
NO. TEETH	96T
PITCH DIA.	Ø 12.00
DIAMETRAL PITCH	8
PRESSURE ANGLE	20 DEGREES

KEYSEAT .12–.25

KEYSEAT .12–.25

NOTE:
ALLOWABLE HP = 10.3
(BASED ON 1.2 S.F.)

Ø 1.13

Ø .88

Ø 12.25

Ø 12.00

Ø 10.94

Ø 3.00

1.25

1.50

.60

Ø 1.62

.75

Ø 1.70

Ø 2.00

2.25

(g)

Fig. 6-3 (g). Spur gear working drawing (cont.).

calculation or approximation methods. Select the CIRCLE and PHANTOM line-type commands.

10. Create the outside diameter and root diameter in phantom line type as shown in Fig. 6-3(b).

11. Select the CIRCLE-CENTER/DIAMETER and SOLID line-type command.

12. Set the cursor at the large gear center (point 5 in Fig. 6-3[a]).

13. Key in 1.12 and press return. The 1.12 diameter hole for the shaft will be created.

14. Use ZOOM IN, SOLID LINE, and COORDINATE input commands to create the .12- × .25-in. keyway (points 1, 2, 3, and 4 in Fig. 6-3[c]). The solid circle serves as a construction guide.

15. Use ZOOM, SNAP OFF, and 3 POINT arc commands. Set the cursor at points 1, 2, and 4, respectively. As an alternative, a BREAK (CIRCLE) option may be used.

16. Select ERASE. Delete the full circle drawn in step 13. The result will appear as shown in Fig. 6-3(d).

17. Using the appropriate commands, continue to layout the large gear. Use your judgment for dimensions not given. The finished gear will appear as shown in Fig. 6-3(e). Gear teeth are not normally drawn on a working drawing. Since their sizes have been standardized, this would be an inefficient, time-consuming process. CADD, however, would significantly diminish this chore. That is, create one tooth and use ARRAY-POLAR for the others.

18. Next, create the small gear. Begin 7.00 in. from the center of the large gear. This will be the center of the small gear since the pitch radii (12/2 + 2/2) must be tangent. Continue to lay out the gear in the same manner as previously described. The result will appear as shown (rotated 90° for convenience) in Fig. 6-3(f).

19. Use the TEXT command to specify the cutting data. A finished working drawing of two gears in mesh is shown (rotated 90° for convenience) in Fig. 6-3(g). Shafts and keys may be designed and added to this project, as required.

Bevel Gears

Bevel gear working drawings may be created in a manner similar to spur gears. Gather the necessary size data from Appendix 10. Use your judgment for sizes not given. Figure 6-4 illustrates examples of finished bevel gear working drawings.

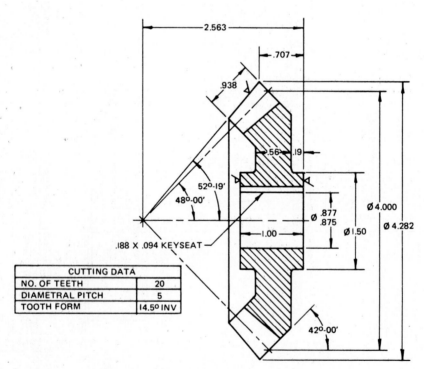

Fig. 6-4 (a). *Working drawing of a bevel gear.*

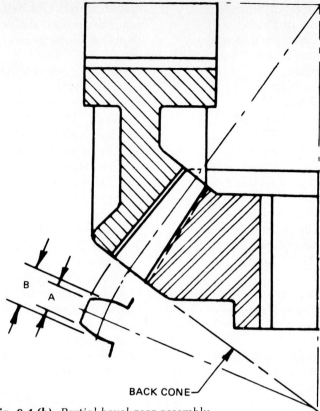

Fig. 6-4 (b). *Partial bevel gear assembly.*

ASSIGNMENTS

1. Create spur gear working drawings to meet the specifications of the following problem(s), as assigned by your instructor:

   a. Single spur gear
      Pressure angle—20°
      PD—6.00
      DP—6
      Face width—2.00
      Web—.40 thick
      Six cored holes
      Shaft—ϕ1.06
      Hub—ϕ3.00 × 1.25 projection

   b. Single spur gear
      Pressure angle—20°
      N—50
      DP—6
      Face width—2.00
      Shaft—ϕ1.18
      Hub—3.50 × 1.25 projection
      6 spokes at .60 in. thick, 1.50 in. wide tapered to 1.10 in. wide.
      Use the POLAR ARRAY command to develop the spokes.

   c. Meshing spur gears
      Pressure angle—20°
      Center to center distance—6.00

	PINION	GEAR
N	24	36
face width	1.10	1.10
shaft	ϕ1.10	ϕ1.25
web		.40
hub		ϕ2.10
hub project		1.50

   Determine the allowable hp of the drive if the pinion rotates at 900 rpm.

2. Given center-to-center distance between gear and pinion of 10 in., gear rpm 400, pinion rpm 1600, 20° angle, calculate D of gear and pinion. Select a pitch, and determine number of teeth and gear and pinion. Prepare a working drawing of two gears in mesh. Show two views. Add suitable keyways, and use your judgment for dimensions not given. Include cutting data for each gear, and determine the allowable horsepower.

3. The spindle subassembly shown in Fig. 6-5 is part of the drill press, Fig. 11-3. Tapered roller bearings will be pressed on the spindle shaft and contained by a cylindrical housing as shown. Foreign objects will be prevented from dropping in the housing by a top plate threaded onto the spindle shaft with clearance between it and the housing. The top plate is locked to the shaft to prevent it from backing off. One side of the housing will be milled to allow attachment of the rack. Since the

rack and pinion are manually operated, forces are low; thus a light-pitch series may be used.

    a. Prepare a subassembly drawing of the rack and pinion spindle shown in Fig. 6-5. Select bearings based on spindle shaft sizes.

    b. Prepare an items list. Select standard manufactured parts for all purchased items.

4. Determine the allowable horsepower for the system shown in Fig. 6-6. Select appropriate bearings based on the maximum allowable horsepower. Prepare an assembly drawing of the drive system.

5. Repeat assignment 4 for the system shown in Fig. 6-7.

6. Repeat assignment 4 for the system in Fig. 6-8.

7. a. From the information shown in Fig. 6-9, design the unit for a 2:1 speed reduction and able to withstand 3 hp at 900 rpm. Use a service factor of .9 for selecting gears.

    b. Prepare a subassembly drawing. Include input and output shafts to their proper length. The output shaft requires an extension of 4 in. from one end of the casting.

    c. Prepare an items list of materials specifying all standard components by catalog number and manufacturer. Use bronze journal bearings on input and output shafts. Include oil seals, keys, retaining rings, setscrews, and a standard 3 hp motor.

    d. Determine critical dimensions required to prepare the casting drawing.

8. A combined belt and gear drive system is shown in Fig. 6-10. The design specifications are 3-hp, 1160-rpm motor operating under normal conditions. The belt drive transmits 1.00 hp to shaft 2 (S_2). The desired S_2 output speed is 500 rpm at 20 in. center to center. The gear drive transmits 2.00 hp to shaft 3 (S_3). The desired S_3 output speed is 350 rpm.

    a. Design and select:
        Belt and pulley drive
        Spur gear drive
        Shafts: S_1 at 3 hp, S_2 at 1 hp, S_3 at 2 hp
        Keys, bearings, coupling
        Present all data and calculations.

    b. Prepare an assembly drawing including an items list.

    c. Determine the system allowable horsepower by listing that of each component.

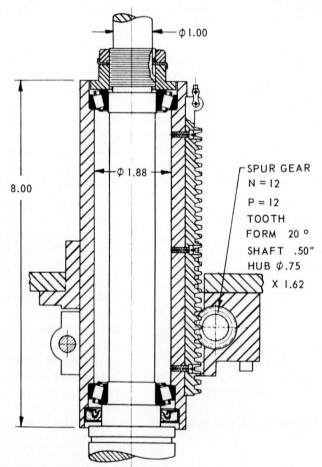

Fig. 6-5. *Drill machine spindle.*

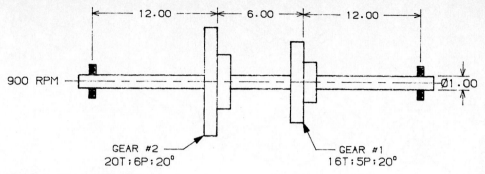

Fig. 6-6. *Gear problem.*

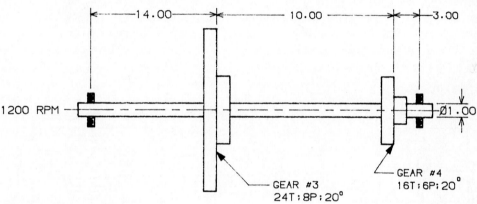

Fig. 6-7. *Gear problem.*

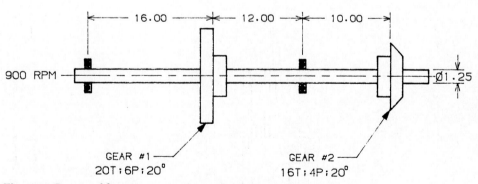

Fig. 6-8. *Gear problem.*

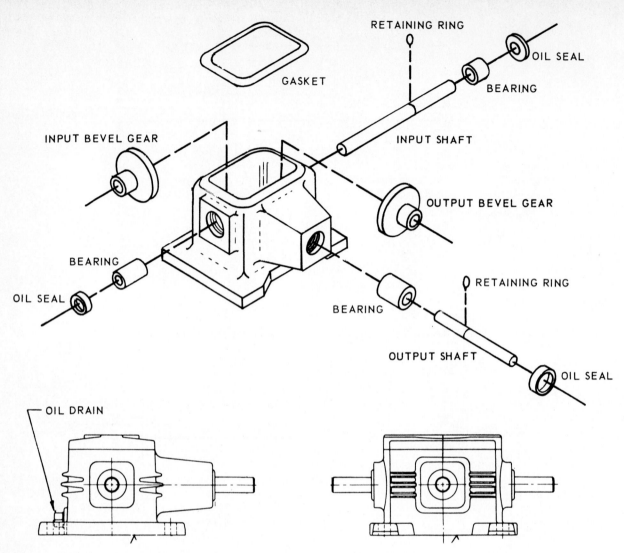

Fig. 6-9. *Bevel gear drive assembly.*

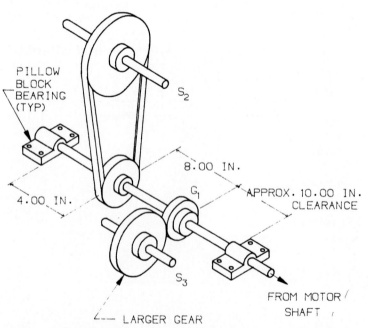

Fig.6-10. *Combined belt and gear drive system.*

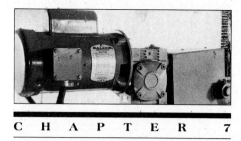

C H A P T E R 7

GEAR REDUCER

DESIGN THEORY

Speed reduction is common practice in industrial machinery. Standard motors rotate at high speeds (for example, 1750 rpm). Output speeds of machines such as conveyors, however, may be only a small percentage of that value. For reductions greater than approximately 6:1, a single pair drive component (spur or bevel gear, pulley, sprocket) cannot be used. Two or more pairs will be required to attain a high reduction. This is referred to as a *compound train*. Two pairs of meshing spur gears may be used for reductions between 6:1 and 36:1. For a speed reduction of 16:1, for example, two 4:1 reductions could be used. If the ratio is greater than 36:1, use three sets of meshing gears.

This chapter will analyze compound trains for spur gears.

Speed Ratio for Compound Gear Trains

The speed ratio for gear trains can be determined simply from the concept that the rpm and the number of teeth are inversely related.

$$\text{Speed ratio} = \text{output speed/input speed} = N_1/N_2$$

7-1

Where N_1 and N_2 are the numbers of teeth on the two gears in mesh.

This can be extended to compound (more than a single mesh) trains.

$$\text{Output speed/input speed} = N_1/N_2 \times N_3/N_4 \qquad \text{7-2}$$

DESIGN APPLICATION

EXAMPLE 1

Step 1
The speed ratio (or train value, TV) for the gear train shown in Fig. 7-1 is

$$20/60 \times 20/50 = 2/15 \text{ or } 2:15$$

If gear N_1 is rotating at 1750 rpm, output gear N_4 will be rotating at 233 rpm in the same direction. The allowable horsepower for this train can be determined by considering each pair independently.

Step 2
If the first set of gears are 20° PA, 8P, the allowable horsepower at 1750 rpm is approximately 15.0 as determined from Fig. 6-2. For the second set of 8-pitch gears at 583 rpm the allowable horsepower is 9.0. Consequently, the allowable horsepower for the train is the lowest value of all meshes, in this case 9.0.

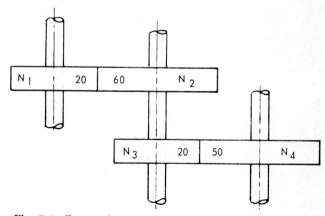

Fig. 7-1. Gear train.

EXAMPLE 2

If 10-pitch gears are used for the first set (in Example 1), the allowable horsepower becomes 10.0. The allowable horsepower for the train remains unchanged because the second set is still the limiting factor. The first set is smaller than before, and the allowable hp of the train is the same. This is considered a better design than the one in Example 1.

It is common to use smaller- or lighter-pitch gears for the first set of a compound train. This is due to the higher rpm of the input shaft. Less torque is required to transmit horsepower at a higher rpm. Thus the drive components need not be as heavy as a low rpm application.

EXAMPLE 3

Two sets of meshing gears are used, with one having a 3:1 reduction and the other a 4:1 reduction. With a 1200-rpm input, the intermediate shaft holding gears 2 and 3 may either rotate at 400 rpm or 300 rpm. Using the 3:1 reduction pair first is desirable since less torque is needed to transmit the horsepower at the faster speed of 400 rpm. The gears, shafts, and keys required will be smaller than they would be at 300 rpm.

EXAMPLE 4

Very often in a compound train only the input and output speeds are known. This requires using computer or trial-and-error methods to determine the number of gear teeth. If the desired speed ratio exceeds 6:1, it is recommended that two sets of meshing spur gears be used. These can be used for a speed change up to

$$1/6 \times 1/6 = 1/36 \text{ maximum}$$

Suppose a speed reduction of 9:1 is desired. This is a simple problem since the ratio can be evenly split into

$$1/3 \times 1/3 = 1/9$$

Multiple solutions exist for numbers of teeth: a 12 and 36 pair, a 15 and 45 pair, and so on. The exact sets used will depend on horsepower requirements and available standard gears.

EXAMPLE 5

A speed reduction of 6.93:1 is desired. The problem is no longer trivial since we are dealing with prime numbers. If the trial-and-error method is used, multiply numerator and denominator by the multiple of the two smallest standard gears desired. If this does not result in a solution, then try the next smallest standard gears.

For trial 1, use 12 tooth gears for N_1 and N_3.

$$1/6.96 \times (12 \times 12)/(12 \times 12) = 144/1002 =$$
$$12 \times 12/(\text{Denominator does not divide into two whole numbers.})$$

For trial 2, try 15 and 20 teeth for N_1 and N_3.

$$1/6.96 \times (15 \times 20)/(15 \times 20) =$$
$$300/2088 = (15 \times 20)/(36 \times 58)$$

However, 58 teeth is not a standard gear. Trial 3 is a variation of trial 2.

$$1/6.96 \times (15 \times 20)/(15 \times 20) =$$
$$300/2088 = (15 \times 20)/(24 \times 87)$$

Eighty-seven teeth is not a standard gear. For trial 4, try 20 teeth for N_1 and N_3.

$$1/6.96 \times (20 \times 20)/(20 \times 20) =$$
$$400/2784 = (20 \times 20)/(42 \times 66) \text{ or } (20 \times 20)/(48 \times 58)$$
$$\text{or } (20 \times 20)/(50 \times 56)$$

These combinations will work since these gears are standard stock items in most pitches (Appendix 9). The speed ratio, while not exact, is very close in each case. Bear in mind that horsepower requirements have been disregarded and pitch is not a factor at this point. The next step is to match the two sets of meshing gears to minimize wear. For the example given, matching the sets is not necessary since N_1 and N_3 have the same number of teeth. If they did not, then it would be best to match the lowest numerator with the lowest denominator.

CADD DRAWING

1. The meshing spur gear assembly $(20 \times 20)/(50 \times 56)$ designed in Example 5, may be drawn. First select an appropriate pitch and pressure angle for each set of gears. If the train must be capable of safely transmitting 7.5 hp at a motor speed of 1800 rpm select:

   10 pitch, 20° PA, 20- and 50-teeth gears 1 and 2 (approximate allowable hp = 9.0 at 1800 rpm)
   8 pitch, 20° PA, 20- and 56-tooth gears 3 and 4 (approximate allowable hp = 9.0 at 720 rpm)
   The critical size for each gear may be found in Appendix 9.

2. Begin a B size, full-scale drawing.

3. Select the CENTER line type. Create vertical centerlines for each of the three shafts. The first centerline is 3.50 in. (1.00- + 2.50-in. pitch radius) from the second. The second is 4.75 in. (1.25-

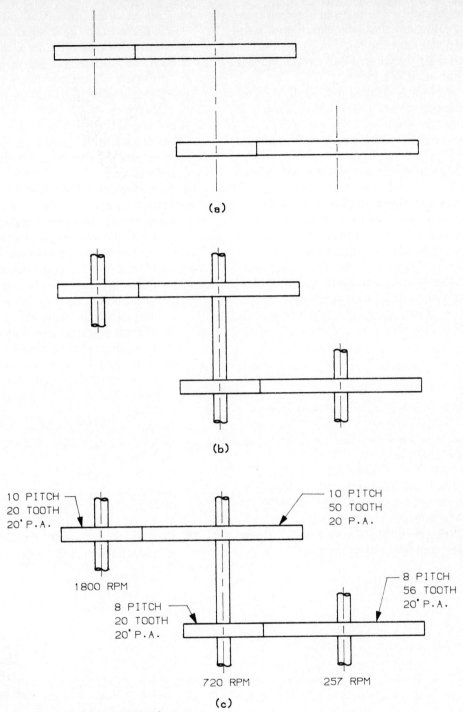

(a)

(b)

10 PITCH
20 TOOTH
20° P.A.

10 PITCH
50 TOOTH
20 P.A.

1800 RPM

8 PITCH
56 TOOTH
20° P.A.

8 PITCH
20 TOOTH
20° P.A.

720 RPM   257 RPM

(c)

***Fig. 7-2**. Gear train CADD drawing.*

+ 3.50-in. pitch radius) from the third.
4. Select LINE. Create the outline of each meshing gear using the appropriate pitch diameter and face width. The result is shown in Fig. 7-2(a).
5. Create the shafts based upon the ϕ1.00-in. bore in each gear.
6. Use (S)PLINE to break the end of each shaft as shown in Fig. 7-2(b).
7. Select LEADER and TEXT. Add the information for each gear and the shaft speeds as shown in Fig. 7-2(c).

8. Include additional information such as the allowable horsepower of the shafts and sets of spur gears.

ASSIGNMENTS

1. The input speed at shaft (a) of Fig. 7-3(a) is 1800 rpm.
   a. Determine the speed reduction for 6 pitch, 20° pressure angle gears.

b. Find the speed and torque at each shaft and determine the allowable horsepower of the compound train (gears and shafts).
c. Create an assembly drawing of the compound gear train including shafts and keys.
2. Gears 1 and 2 in Fig. 7-3(b) have a pitch of 8 and 20° pressure angle. Gears 3 and 4 have a pitch of 6 and 20° pressure angle.
   a. Determine the allowable horsepower of the train.
   b. Create an assembly drawing of the compound train.
3. Find the speed and direction of rotation of gear 7 in Fig. 7-4 if shaft (a) rotates counterclockwise at 600 rpm. Determine the speed reduction ratio. Helical gears 4 and 5 mesh at right angles.
4. a. Determine, as closely as possible, a gear train shown in Fig. 7-5 that will yield any of the following (as assigned by instructor) speed reductions. Use standard gears. The smallest gear to be used is a 12 tooth.
   1:19
   1:17.1
   1:15.3
   1:14.7
   Determine a workable train and identify the actual ratio that is used (ideally it will be exact). Match the gears for wear, placing them in the appropriate position.
   b. Using Fig. 6-2, determine the allowable horsepower of the train. The motor to be used will be a standard 1750 rpm. Use 8- or 10-pitch, 20° pressure angle gears.
   c. Determine the required sizes for the shafts and square keys. Match these to the gear bores.

d. Present calculations and rationale to support each aspect of the design.
e. Prepare an assembly drawing.
f. Prepare an items list. Select standard manufacturer parts including the motor and coupling.
g. Prepare detail shaft and casting drawings.
5. Use a face-mounted .50-hp, 1750-rpm motor to drive an output load through shaft 4 of the motor drive assembly sketch shown in Fig. 7-6. Neglect system inefficiency. Output shaft 5 is used to drive a dial indicator.
   a. Design gears, shafts, and keys to meet system requirements. Complete the tables in Fig. 7-6, and answer the following questions:
      1. How many seconds does it take for the load-control output shaft to rotate one revolution?
      2. How many seconds does it take for the dial indicator shaft to rotate one revolution?
      3. What is the speed reduction between motor and output load shaft?
      4. What is the speed reduction between motor and dial indicator shaft?
      5. If the load control shaft rotates 1800°, how many revolutions has the dial indicator rotated? How many revolutions has the motor rotated?
   b. Prepare a CADD assembly drawing. Determine overall required sizes for the system brackets and supports.
   c. Prepare an items list and include it on the CADD-generated assembly. Use standard manufactured parts for the gears, motor, journal bearings, and hardware.
   d. Prepare a necessary detail drawing of all nonpurchased components.

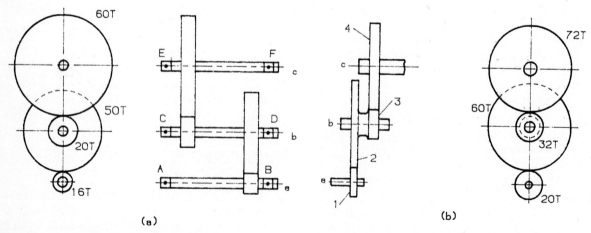

(a)   (b)

Fig. 7-3 (a and b). *Gear train problems.*

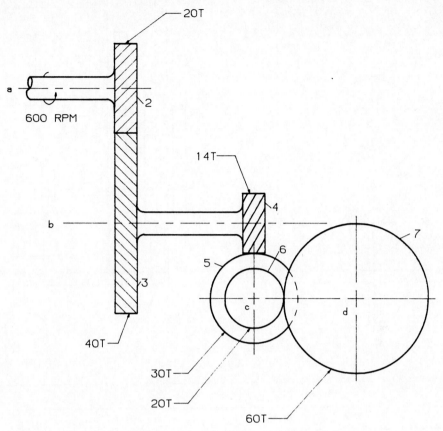

Fig. 7-4. Gear train problem.

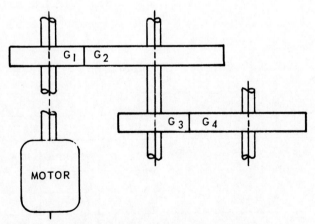

Fig. 7-5. Compound spur gear reducer assembly.

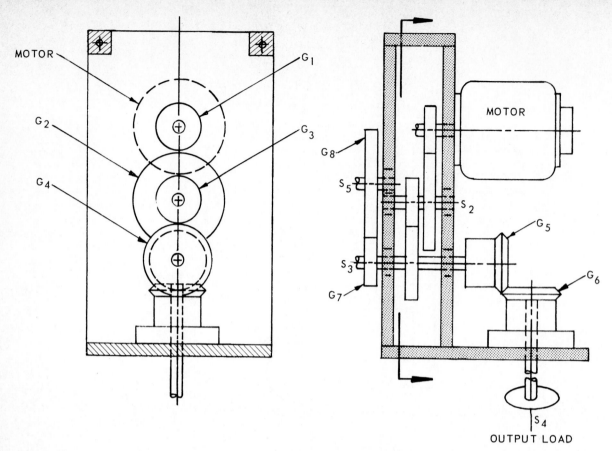

MOTOR DRIVE—GEAR DATA							
GEAR	SPEED RATIO	NUMBER OF TEETH	PITCH	ALLOWABLE HP	BORE	HUB DIA	HUB LENGTH
1	1:4						
2							
3	1:5						
4							
5	1:1						
6							
7		14 OR 15		MIN REQD			
8		70 OR 72					

SHAFT DATA				
SHAFT NO	RPM	REQD DIA	ACT DIA	ALLOWABLE HP
1				
2				
3				
4				
5				MIN REQD

KEY DATA				
KEY NO	ACT RPM	SQ SIZE	REQD LENGTH	ACT LENGTH
1			MIN REQD	
2				
3				
4				
5				
6				
7				
8				

Fig. 7-6. Motor drive assembly.

MECHANICAL DESIGN USING CADD

CHAIN

DESIGN THEORY

A chain drive system will likely be used for applications that require transmitting heavy materials at virtually any horsepower. In addition, since shaft center distances are relatively unrestricted, these may be transmitted over long distances. There are many different types of chain drives. Some include:

- Roller
- Silent
- Pintle
- Engineering steel
- Cast

Of all the types available, roller chain is the most widely used, because of its versatility. This chapter will concentrate on the application of roller chain. A typical roller chain is shown in Fig. 8-1.

Roller chain numeric designation is standardized. The first or first two numbers identify the pitch of the chain. This is the center-to-center distance between pins. The number specifies how many eighths of an inch the pitch is. The last number indicates whether the pins have rollers, are lightweight, or are rollerless. Small pitch links are rollerless. This is due to a space limitation. The use of rollers would not provide enough room for the sprocket teeth to properly mesh. Some typical examples are:

35—.38 (3/8) pitch rollerless (5) type
50—.62 (5/8) pitch roller (0) type
100—1.25 (10/8) pitch roller (0) type

If you see a hyphen with a number to the right of the standard chain number, it will refer to the number of strands.

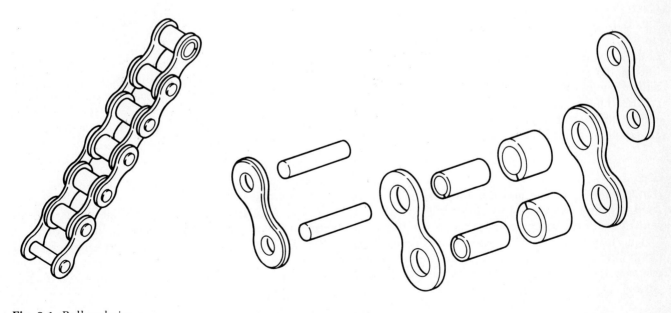

Fig. 8-1. *Roller chain.*

	TYPE OF INPUT POWER		
Type of Driven Load	Internal Combustion Engine with Hydraulic Drive	Electric Motor or Turbine	Internal Combustion Engine with Mechanical Drive
Smooth	1.0	1.0	1.2
Moderate Shock	1.2	1.3	1.4
Heavy Shock	1.4	1.5	1.7

(a)

SMOOTH LOAD	MODERATE SHOCK LOAD	HEAVY SHOCK LOAD
Conveyors—Uniformly loaded Fans—Centrifugal and light, small diameter Line shafts—Light service Machines—All types with uniform nonreversing loads	Conveyors—Heavy duty and NOT uniformly loaded Cranes and hoists—Medium duty, skip hoists Laundry industry—Washers, tumblers Line shafts—Heavy service Machine tools—Main and auxiliary drives Machine—All types with moderate shock and non-reversing loads Screens—Rotary (stone or gravel)	Conveyors—Reciprocating and shaker Cranes and hoists—heavy duty Hammer mills Machine tools—Punch press, shears, plate planners Machines—All types with severe impact shock loads or speed variation, and reversing service

(b)

Fig. 8-2. (a) Service factors for roller chain drives; (b) load classifications.

Design Procedure

Step 1

As with other mechanical drive components, the service factor must be determined when you begin the design. This may be taken from the table in Fig. 8-2.

$$\text{Design hp} = \text{motor hp} \times \text{service factor} \quad \text{8-1}$$

Step 2

Next, determine the tentative chain pitch. Depending on the number of strands, use the appropriate column at the left side of Fig. 8-3. Find the intersection of the horsepower to be transmitted and the speed of the smaller sprocket. The intersection point will fall within one of the inclined bands in Fig. 8-3. This indicates the tentative pitch. You may end up varying somewhat from this recommendation; however, it is an excellent point from which to begin.

Step 3

If more than a single strand as shown in Fig. 8-4(a) is used, a *strand factor* must be applied. This is due to multiple strands being stronger than a single one.

Thus to transmit the same power, the multiple-strand chain and sprocket are smaller. Two strands, however, are not twice as strong as one strand. The strength (strand) factor for multiple strands is:

NUMBER OF STRANDS	STRAND FACTOR
2	1.7
3	2.5
4	3.3

Thus when designing under these conditions

$$\text{Design hp/strand} = \text{design hp/strand factor} \quad \text{8-2}$$

Step 4

The number of pitches (links) in the approximate center distance must be determined. This is determined by

$$\text{Number of pitches} = \text{center distance (in.)}/\text{pitch (in.)} \quad \text{8-3}$$

The optimum value for the number of pitches is between 30 and 50 links. As this number increases, so does the chain weight. The result is sagging chain. If the number exceeds 80, the sag may become excessive. Thus keep each design under this value. If you must exceed 80, provide support.

Step 5

Select the small sprocket on the basis of tentative chain pitch. Use the table in Appendix 11. If, for example, a 60 pitch was recommended, use the .75 in table. The required number of teeth will be found in the left vertical column. Find a horsepower value meeting or exceeding the design horsepower from the rpm column. "Finger-follow" horizontally to the left for the number of teeth to specify.

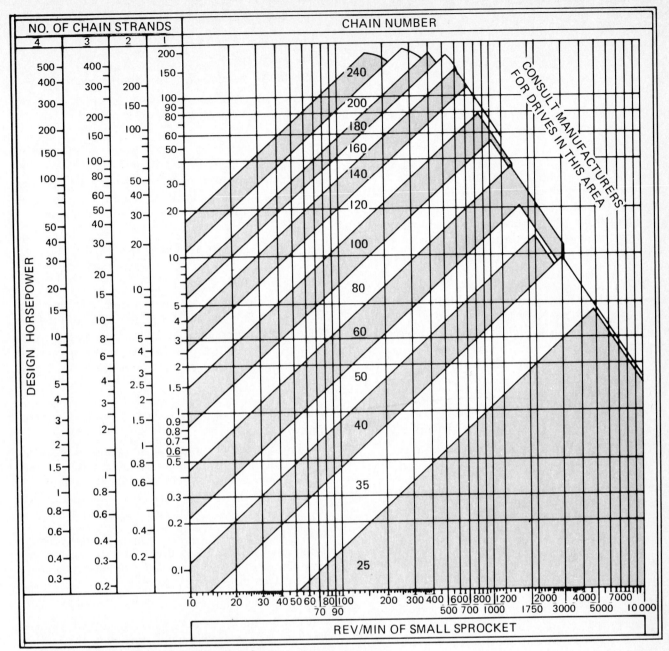

Fig. 8-3. *Roller chain pitch selection chart.*

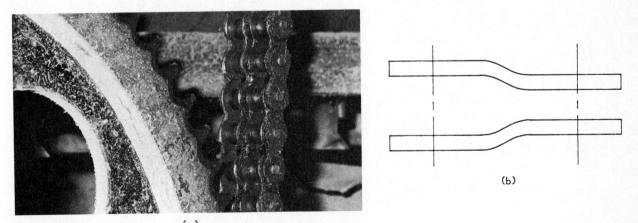

(a)

(b)

Fig. 8-4. *(a) Multiple-strand chain; (b) offset link.*

CHAIN

Step 6

The required size for the output sprocket is easily determined. As in gears, the output speed is continuous. It is inversely proportional to the number of teeth in the sprockets. Thus

Output speed (rpm)/input speed (rpm) = number of teeth for input sprocket/number of teeth for output sprocket   8-4

When solving the equation, you may find the result is not a whole number. Round it to the closest whole number. The smaller pitch sprockets are available in practically any number of teeth. Refer to Appendix 12. If you need to round, the output rpm will change. Recalculate Eq. 8-4 to find the actual output rpm.

Step 7

Next, the chain length must be determined. *Caution:* All values will be expressed in number of pitches (links), not inches as with belt length determination. If a center distance is in inches, it must first be converted to pitches. Use Eq. 8-3 for this value.

To determine the chain length, calculate the following equation:

$$l = 2C + (N + n)/2 + (N - n)^2/4\pi^2C \qquad 8\text{-}5$$

where l = chain length (number of links)

C = center distance (pitches)

N = number of teeth in large sprocket

n = number of teeth in small sprocket

π = 3.14

After substituting the values, round the result to an even integer. A chain will only couple with an even number of links. If the center distance is critical, however, you may have to round to the closest integer. The result may be an odd number of links. If this is the case, an offset link as shown in Fig. 8-4(b) must be used to complete the endless loop.

You may wish to recalculate the exact center distance. This can be done by re-solving Eq. 8-5 for C. If you try to do this, however, you will see that it is not an easy task since it is a polynomial $(C^2 + C + K)$. The quadratic formula may be used to obtain the result. (Yes, there is a real-life practical application to the quadratic formula.) To simplify matters, you can re-solve Eq. 8-5 disregarding the end term at the right. For low-speed ratios and long center distances, the approximation will be very close.

Step 8

Specify the complete drive system you have designed by the following:

Pitch, number of strands, and length of chain

Pitch and small sprocket number of teeth

Pitch and large sprocket number of teeth

Design horsepower

Allowable horsepower of the chain system

Actual output speed

Actual center distance

If the project extends into the design of other components, use the procedures developed in earlier chapters. For example, design a shaft and key for each sprocket. Check Appendix 12 to be certain that you are not exceeding the maximum allowable bore.

DESIGN APPLICATION

SPECIFICATION

Design a roller chain conveyor system to be powered by a 5-hp, 1200-rpm electric motor. The output shaft is located approximately 22.50 in. away and operates between 378 and 382 rpm. The load on the main shaft is not uniform, placing it in the moderate shock-load category.

SOLUTION

Step 1

Use Fig. 8-2 to determine the service factor. For an electric motor operating a conveyor with uneven loads, the factor is 1.3. The design horsepower is $5.00 \times 1.30 = 6.50$ hp.

Step 2

Select the tentative chain pitch. Use the horsepower rating chart in Fig. 8-3 to find that the suggested selection for a 6.50 design hp and a 1200-rpm sprocket is a No. 40 (.50-in.-pitch) chain.

Step 3

Decide the number of strands the system will require. Since this is a relatively low horsepower application with no special requirements, a single strand will suffice. Thus no strand factor is involved.

Step 4

The number of links in the approximate center distance is

Number of pitches = 22.50/.50 = 45

This result is satisfactory.

Step 5

Select the small sprocket on the basis of tentative chain pitch. By our using the number 40 chain table in Appendix 11 at 1200 rpm, the computed design horsepower of 6.50 is realized with a 19-tooth sprocket. The allowable horsepower is 7.27. This need not be checked again for the larger sprocket since the smaller one is the limiting factor.

Step 6

Select the large sprocket. Since the driver sprocket is to operate at 1200 rpm and the driven sprocket at a minimum of 378 rpm, the maximum speed ratio is 1200/378 = 3.175:1.

Therefore, the large sprocket should have

$$N = n \times \text{input speed/output speed}$$
$$= 12 \times 3.175 = 60.32 \text{ teeth}$$

Since 19- and 60-tooth sprockets are acceptable and standard, use this combination. The actual output rpm becomes

$$\text{rpm}_{out} = (1200 \times 19)/60 = 380 \text{ rpm}$$

Step 7

Determine the chain length in links (pitches).

$$l = 2C + (N + n)/2 + (N - n)^2/4\pi^2C$$

$$2(22.50/.50) + (60 + 19)/2 + (60 - 19)^2/4$$
$$\times 9.85 \times 45 = 130.448$$

Since the center distance is not critical, the chain will better couple with an even number of links. Thus, use 130.

Step 8

The drive system is:

40 pitch × 130 links

40-pitch, 19-tooth sprocket

40-pitch, 60-tooth sprocket

Design horsepower = 6.50

Allowable horsepower = 7.27

Actual output speed = 380 rpm

Actual center distance just under 22.50 in.

CADD DRAWING

Chain drive assemblies may be represented schematically. The drawing for the system designed in the Section "Design Application" is created in a straightforward manner as follows:

1. Boot the system at SETUP a B size, HALF scale drawing.
2. Select CIRCLE. To create the sprockets in a schematic representation, you must first determine the pitch diameters. These may be found in Appendix 12(a) or in any standard manufacturer's parts catalog. The diameters may also be mathematically calculated by

Circumference = number of teeth (links) × pitch
Circumference = π × diameter
Thus, C = 19 × .50 = 9.50 (approximate)
    D = 9.50/3.14 = ϕ3.00 in. (approximate; 3.038 in Appendix 12a)
    C = 60 × .50 = 30.00
    D = 30.00/3.14 = ϕ9.50 in. (approximate; 9.554 in Appendix 12a)

Create the small sprocket using CENTER/DIAMETER.

3. Measure 22.50 in. to the right. You may use a .50-in. grid or relative coordinate for this purpose. Create the large sprocket.
4. Select LINE, PHANTOM, and OBJECT SNAP (TANGENT). Draw two lines representing the chain tension andslack sides. The drawing will appear as shown in Fig. 8-5(a).
5. Select CENTERLINE. Create centerlines through each of the sprocket centers.
6. ZOOM in on each of the sprockets. Draw the bore and key using 3 POINT ARC and LINE. Use the maximum sizes listed in Appendix 12. The result will appear as shown in Fig. 8-5(b).
7. ZOOM in at the lower left of the drawing. Create the title block and items list outlines. If a standard library is not available, use LINE for this purpose. Refer to Appendix 6 for sizes.
8. Select TEXT and .12 HEIGHT. Identify each part as shown in Fig. 8-5(c).
9. Add the additional design information.
10. If you have used the quadratic formula to recalculate the center distance, add it to the drawing

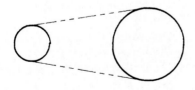

(a)

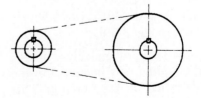

(b)

Fig. 8-5 (a and b). Chain drive system.

3	1	SPROCKET	#40P-60T	STEEL
2	1	SPROCKET	#40P-19T	STEEL
1	130	CHAIN	#40 PITCH	STEEL
ITEM	QUAN.	TITLE	SIZE	MATERIAL
		ITEM LIST		

CHAIN DRIVE SYSTEM	
MARK VOISINET	
DATE:7-4-88	8-A
SCALE: 1.00:2.00	

(c)

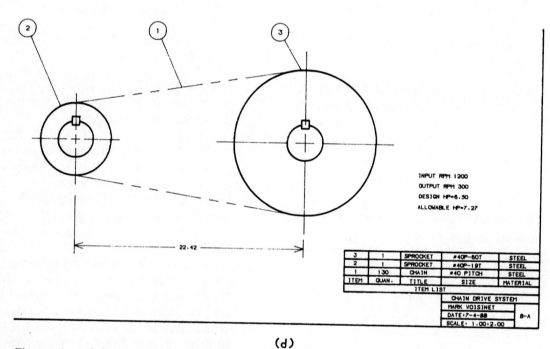

(d)

Fig. 8-5 (c and d). *Chain drive system (cont.).*

using DIMENSION. You may also wish to add an adjustable takeup so that the center may be changed as required. Refer to Appendix 13 for standard takeups. Draw the takeup outline only.

11. Use LEADER, NOTE, and CIRCLE to identify each item as shown in Fig. 8-5(d).

12. If the design is to be more inclusive, shafts, keys, motor, and coupling may be added to the items list and identified on the drawing. This will require an additional (plan) view. Retaining rings and associated hardware may also be used. Refer to Appendix 14 for standard fasteners.

ASSIGNMENTS

1. Refer to Fig. 8-6 for chain drive system specifications.

   a. Determine the large sprocket size and the chain length.

   b. Determine the allowable hp of the system and the actual output rpm.

   c. Create a schematic assembly drawing. Use Fig. 8-5 as a guide.

2. A machine producing moderate shock is to be driven at approximately 40 rpm by a 5-hp, 100-rpm electric motor. The shaft center distance is approximately 36.00 in.

a. Design a single-strand chain drive system with the appropriate sprockets and chain.

b. Prepare an assembly drawing. Place rationale on the drawing with notes to support the design including allowable horsepower and actual output speeds.

c. Prepare an items list. Select standard manufacturer parts for the chain and sprockets. The motor and machine reducer are existing and need not be selected.

3. The same as assignment 2 except a double-strand chain is to be used.

4. Design a chain drive system to meet the following specifications:

20 hp at 1200 rpm input, 1.2 service factor, 22.50 in. center distance, and 756 rpm out.

a. Select suitable chain and sprockets.

b. Design the input and output shafts. Be certain each does not exceed the maximum sprocket bore specified in Appendix 12.

c. Prepare a two-view assembly drawing. Include all design data. Hold each sprocket in position with a pair of retaining rings. Select rings from Appendix 14, according to shaft diameters. Draw the rings as shown in Fig. 2-5(b). Select a coupling to connect the motor shaft to the input shaft.

d. Prepare an items list.

5. The 100-hp, 900-rpm motor shown in Fig. 8-7 drives three machines. Machine A requires 75 hp at 400 rpm and has a center distance of approximately 4 ft 0 in. Machine B requires 15 hp at 500 rpm with the same center distance. Machine C requires 10 hp at 100 rpm and is located approximately 4 ft 0 in. above the intermediate shaft center.

a. Design the four chain and sprocket systems required to drive these machines.

b. Determine the size of the intermediate shaft located approximately 5 ft 0 in. from the motor shaft. Match the shaft diameter to the sprocket bores.

c. Present calculations and rationale to support each aspect of the design.

d. Prepare an assembly drawing.

e. Prepare an items list. Select standard parts for the chains, sprockets, and motor. Hold the sprockets in position with retaining rings (Appendix 14). Machines A, B, and C are existing and need not be selected.

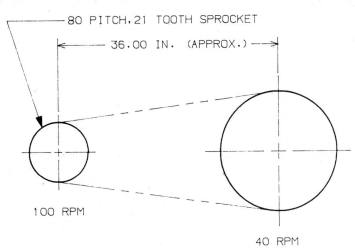

Fig. 8-6. *Chain problem.*

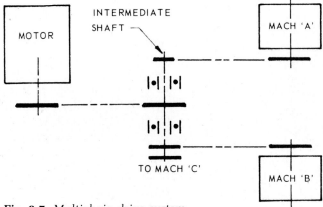

Fig. 8-7. *Multichain drive system.*

CHAIN

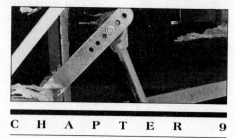

CAM AND LINKAGE

DESIGN THEORY

Cams or linkages are commonly found on machines. These types of components provide an option for a nonconcentric or nonlinear output. The result produces a variable motion. It is possible that an unacceptable velocity or acceleration may exist. The effects of this must be examined. After the analysis, the working drawings may be completed. This will normally not be complicated. Other than the use of (S)PLINE for cam layouts, basic drawing primitives will be applied. Figure 9-1 illustrates a layout of a plate cam profile. It utilizes (S)PLINE with LINE and CIRCLE commands. Of equal importance to a designer is the type of motion produced by the cam as it moves a linkage (follower).

Motion analysis in cam and linkage systems is used to determine velocity and accelerations at critical points of the program. Solutions may be found through the use of graphs and are accomplished by use of the slope concept. The procedure used to accomplish this follows.

Step 1
The slope of any given line is determined by dividing the amount of rise (elevation) by the length it has taken to rise to that height. Other common ways of expressing this are the rise over the run or change in y (Δy) over change in x (Δx). The slope shown in Fig. 9-2 thus is

$$m = \Delta y/\Delta x = (5 - 1)/(2 - 0) = 4/2 = 2$$

If the inclined straight line descends to the right rather than rising, the slope becomes a negative value. For example, the slope of the given line shown in Fig. 9-3 is

$$m = \Delta y/\Delta x = (1 - 5)/(2 - 0) = -4/2 = -2$$

Step 2
Cam profile displacement will usually not be linear. For curved lines the slope can be determined only at particular points, since it is constantly changing. Construct a line tangent to the point where the slope is desired, as shown in Fig. 9-4. Construct a convenient triangle and calculate the slope as before. For this example the slope of the curve at the desired point is

$$m = \Delta y/\Delta x = (4.70 - 1)/(3.50 - 1) = 3.70/2.50 = 1.48$$

Step 3
Consider units to be associated with the numbers in Fig. 9-2. For example, if the first given line rose 4 mi in 2 hours, the slope becomes 2 mi/h, a unit of velocity. If inches and seconds are used as units in Fig. 9-4 for the vertical and horizontal axes, respectively, the velocity of the curve at that point becomes 1.48 in./s.

Step 4
Take slopes at various places on the upper curve as shown in Fig. 9-5. If the curve is a displacement diagram (for example, the motion of a cam, or follower),

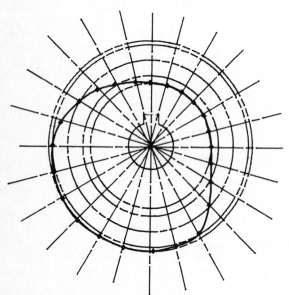

Fig. 9-1. Plate cam.

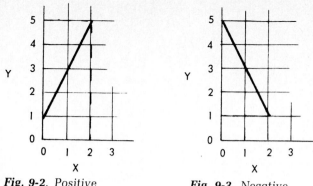

Fig. 9-2. *Positive slope.*

Fig. 9-3. *Negative slope.*

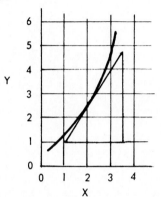

Fig. 9-4. *Changing slope.*

Notice by the tangent lines taken in the displacement diagram (Fig. 9-5) that the slope is increasing from 0 to 3s. Consequently, the velocity of the follower is likewise increasing. There is a constant negative slope for the last portion of the curve; therefore, a constant negative velocity is developed during that part of the cam follower return. It will be equal to −3 in./s.

Step 5

Once the velocity curve has been plotted, repeat Steps 1, 2, 3, and 4 from this new derived (plotted) curve as shown in Fig. 9-5. The slopes taken at A, B, and C will be constant. When inches per second are divided by seconds, the new unit becomes inches per second squared. These are units of acceleration. Plot these values directly below the velocity diagram curve as illustrated in Fig. 9-6. Since the results of the final curve yield a horizontal line, the cam follower is moving with constant acceleration from 0 to 3 s. From 3 to 4 s the velocity is constant. The acceleration during this period will be zero since it is the slope of a horizontal line. A potential problem area is seen at the 3-s position. This is because the acceleration changes instantaneously (vertical line).

When the velocity and acceleration curves are derived, each point must correspond. The slope of the velocity curve at time 1 s is the acceleration at time 1 s. Notice how the shapes of the curves change during the derivation process. The curve portion of the

the slopes will be the velocities of the cam follower at those points. Since you are dividing inches by seconds, the resulting units are inches/second. These are units of velocity. Plot the values on a new curve called the *velocity diagram* directly below the displacement diagram.

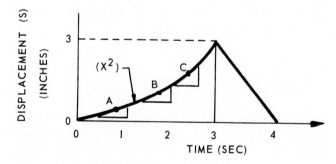

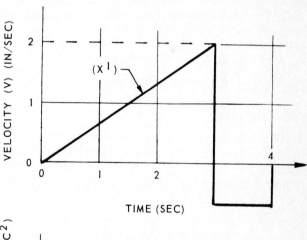

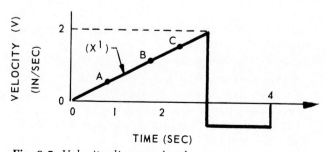

Fig. 9-5. *Velocity diagram development.*

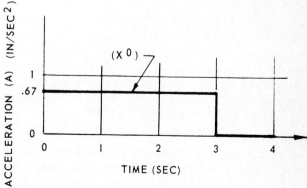

Fig. 9-6. *Acceleration diagram development.*

displacement diagram becomes a linear inclined line for that portion on the velocity diagram from taking the slopes. This inclined line then becomes a horizontal line on the acceleration diagram. It represents a normal course of mathematical events, since the curves are actually changing from X^2 (inches) to X^1 (inches per second) to X^0 (inches per second squared). The acceleration is constant, and this type of motion is called *parabolic* or *uniformly accelerated and retarded motion* (UARM). If the cam displacement was a cycloidal motion (X^3), the corresponding velocity becomes parabolic (X^2), and the corresponding acceleration becomes a constant increase or decrease (X^1).

When determining locations to take slopes on any curve, choose more places where the curve is changing rapidly. Also, try to find a point of inflection or transition. These are places where the slope will be the greatest, and they are important for analyzing results.

Step 6
The results of the curve can now be analyzed. This is normally done by the engineer who looks for critical points during the program. Abrupt changes are important since a discontinuous curve means excessive jerks. Stated another way, if the follower is changing acceleration during zero time it will "slam" against the cam surface, and excessive wear may result at that point. This occurs at the 3-s position. Also, the maximum values, both positive and negative, are looked at to determine if they are too rapid for the system.

DESIGN APPLICATION

SPECIFICATION
A plate cam raises a ϕ.50-in. roller follower 3.00 in. in 180° with a parabolic motion. It has no rise (dwell) for 60° and drops with a parabolic (UARM) motion for the remainder. Determine a motion analysis if the cam rotates at 120 rpm.

Step 1
Lay out the displacement diagram. Use the division ratio of 1:3:5:5:3:1 for the parabolic motion. Refer to a cam chapter in any traditional mechanical design text for a detailed explanation. Add the appropriate units. Since the cam rotates at 120 rpm, one revolution will take .50 s. The displacement will appear as shown in the top diagram of Fig. 9-7.

Step 2
Take slope readings at each 30° increment. The velocity (slope) increases from zero to maximum (25 in./s) in 90° and returns to zero at 180°. A dwell period has zero slope. The return portion has a neg-

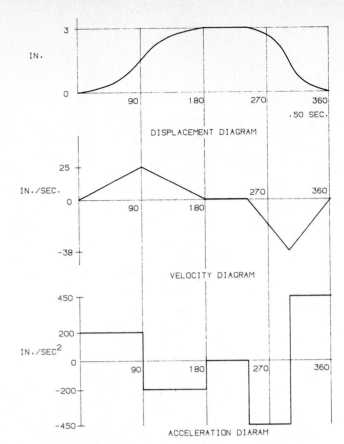

Fig. 9-7. *Motion analysis.*

ative maximum velocity (-38 in./s) at 300°. Plot each velocity under the displacement diagram in Fig. 9-7.

Step 3
Take slope readings for the four inclined velocity lines. Plot each value as shown by the acceleration diagram in Fig. 9-7. Potential problem areas are seen at 0, 90, 180, 240, and 300°.

CADD DRAWING

Create a working drawing of the plate cam analyzed in the Section "Design Application." It has a prime (base) circle of 4.00 in., plate thickness of .38 in., bore of 1.00 with a .12 × .25 keyway. The procedure is as follows:

1. SETUP the system with a full-scale, C size drawing.
2. Lay out the centerline position for the roller follower. This is the distance away from the prime circle equal to the height taken from the displacement diagram plus the roller radius. At 90°, for example, it will be 1.50 in. + .25 in. = 1.75 in. Do this for each 30° increment.
3. Using ARC CONSTRUCTION LINE, create a ϕ.25-in.

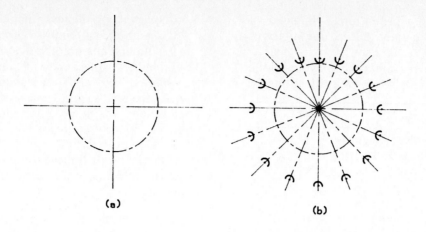

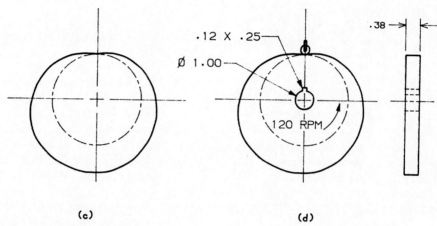

Fig. 9-8 (a–d). *Cam layout.*

arc about each point as illustrated in Fig. 9-8(b). If this is not an option on your system, draw the arcs and erase each one after use.

4. Use the (S)PLINE option to create the cam surface profile as shown in Fig. 9-8(c).

5. Add the bore, keyway, second view, dimensions, and follower as shown in Fig. 9-8(d).

ASSIGNMENTS

1. a. A plate cam raises a .44-in. diameter roller .88 in. with a parabolic (UARM) motion in 120°, dwells 60°, and drops with UARM motion for the remainder. Lay out the displacement diagram for this motion.

   b. If the cam rotates at 720 rpm, determine values for the velocity at 30° increments. Lay out the derived velocity curve directly under the velocity diagram.

   c. Determine the acceleration values at 30° increments from the derived velocity diagram. Lay

out the derived acceleration curve directly under the velocity diagram.

   d. Determine the required shaft and key sizes for the cam if it is powered by a 7.50-hp motor.

   e. Prepare a cam layout drawing using the following information: prime circle = 3.25 in., plate thickness = .28 in., hub = 1.62 in. diameter × .50 in. extension.

2. a. Lay out the displacement diagram that will give a .62-in.-diameter follower the following motion: Dwell for 30°, rise 1.00 in. with UARM motion in 90°, dwell for 45°, rise .75 in. with UARM motion in 60°, drop with modified uniform motion for the remainder. This motion is a straight line with an arc at the beginning and end. Use a radius size of one-half the full rise height. Prime circle = 3.75 in., plate thickness = .43 in., hub diameter = 1.75 in., and hub length = 1.06 in. The cam is driven by a 10-hp, 1160-rpm motor.

   b. Repeat procedures similar to assignment 1, as assigned by the instructor.

3. This is the same as assignment 1, except lay out

a face cam having a prime circle diameter of 6.25 in., cam thickness of .88 in., and a groove depth of .50 in.

4. Prepare motion diagram and profile layouts of the plate cams from the following specifications:

	CAM A	CAM B	CAM C
Base circle	2.88	5.12	4.50
Roller	1.00	1.12	1.25
Direction of rotation	CCW	CW	CCW
Plate thickness	0.50	0.75	0.50
Hub dia. and extension	1.50 × 0.50	2.00 × 0.50	1.50 × 0.25
Bore	0.75	1.25	1.00
Motion	Rise 2 in. in 180° with UARM. Fall 2 in. in 180° with UARM.	Rise 0.75 in. in 60° with UARM. Rise 0.75 in. in 180° with UARM. Fall 1.50 in. in 120° with modified uniform motion.	Rise 1.00 in. in 90° with UARM. Dwell for 90°. Fall 1.00 in. in 180° with UARM motion.

Draw each cam profile using (S)PLINE. Use the plate cam profile illustration shown in Fig. 9-1 as a format guide.

5. a. Lay out the oscillating unit and lever shown in Fig. 9-9 using the sequence assigned by the instructor. Place the lever horizontally in the dwell down position. Show on the cam displacement diagram, in addition to the follower displacement, the lever displacement taken at the very end of the lever. Use your judgment for dimensions not given.

b. From the displacement diagram determine the shape of the velocity diagram.

6. Lay out the Watt's approximate straight-line mechanism shown in Fig. 9-10, and plot the path taken by point E located midway between points B and C. Plot points every 15°.

7. Lay out the toggle linkage shown in Fig. 9-11, and plot the distance x for every 15° of rotation of point A.

AB = 2.50 in., BC = 1.75 in., BD = 2.25 in.

8. Lay out the pantograph mechanism shown in Fig. 9-12, and plot the path of point T. Plot points every 30°.

AP = 4 in., AB = 2 in., BT = 1 in.

9. a. Lay out the shaper motion diagram shown in Fig. 9-13 for two complete strokes. Take positions every 30° of trunnion rotation starting at position 240°.

b. Lay out the velocity diagram for the shaper if the trunnion rotates at 60 rpm.

c. Lay out the acceleration diagram from the velocity diagram derived in (b).

10. Repeat assignment 9 by changing the 2.50-in. trunnion radius to 4 in.

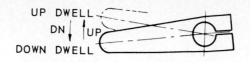

SEQUENCE	UP	UP. DWELL	DOWN	DOWN- DWELL
A	35	30	35	260
B	35	90	35	200
C	45	0	45	270
D	45	60	45	210
E	60	30	60	210
F	60	90	60	150
G	90	0	90	180
H	90	60	90	120

ALL MOVEMENTS ARE UARM
MOTION

LEVER	R IN INCHES
L₂	2.00
L₃	3.00
L₄	4.00
L₅	5.00
L₆	6.00

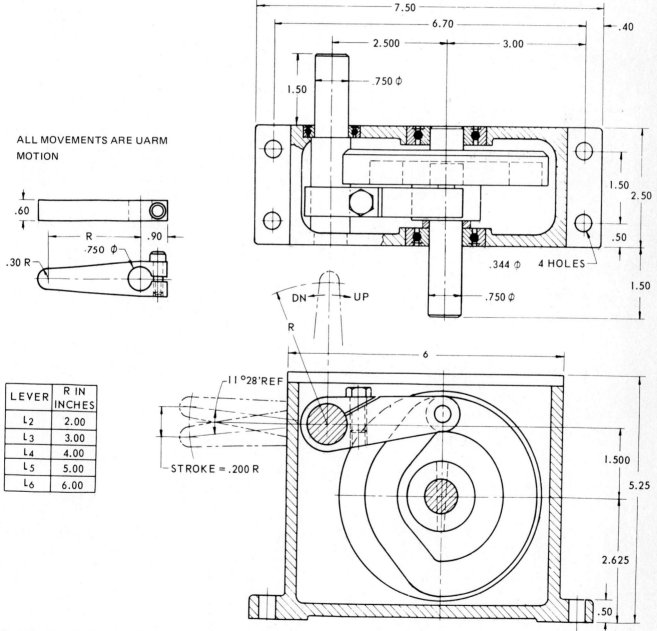

Fig. 9-9. *Oscillating unit (Steiron Cam Co.).*

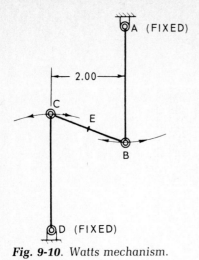

Fig. 9-10. Watts mechanism.

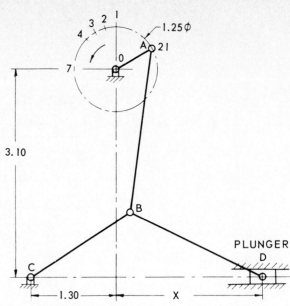

Fig. 9-11. Toggle linkage.

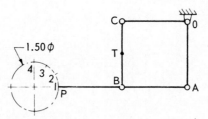

Fig. 9-12. Pantograph mechanism.

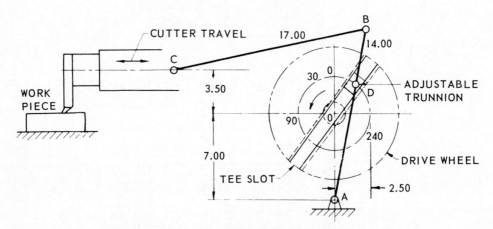

(A) SHAPER SHOWING QUICK RETURN MECHANISM

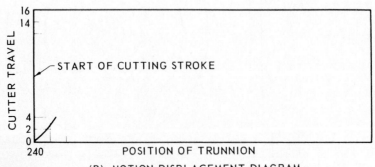

(B) MOTION DISPLACEMENT DIAGRAM

Fig. 9-13 (a and b). Shaper using Whitworth mechanism.

C H A P T E R 10

CABLE

DESIGN THEORY

Cable is widely used in conveying and hoisting equipment. There are many different varieties and types. Three widely used types will be considered in this chapter. They are 6 × 7, 6 × 19, and 6 × 37. The 6 × 7 type means that the cross section of the cable consists of six strands with seven wires per strand as illustrated in Fig. 10-1. It is not very flexible and is used for straight hauling or guy wires. If it must be bent, a sheave with a diameter of at least 42 times the cable diameter must be used.

The 6 × 19 type has six strands with 19 wires per strand. This is somewhat more flexible and is utilized in hoisting equipment. The required sheave diameter is of moderate size, being at least 30 times the cable diameter. The 6 × 37 cable has 37 wires in each of its six strands. It is considered very flexible with a minimum sheave 18 times the diameter of the cable used. Using sheaves of at least this size allows for the effects of bending moments and induced bending stresses caused by wrapping the cable.

The size of cable depends upon cable material and the type of application. Common cable materials and the associated breaking stresses are:

Cast steel, 140,000 psi

Mild plow steel, 160,000 psi

Plow steel, 175,000 psi

Improved plow steel, 200,000 psi

The term *plow steel* is a carryover from the earlier days of a predominately agricultural society. Cables are overdesigned by using a factor of safety. The amount of overdesign varies according to the specific application. Typical factor-of-safety values are:

Guy and tower wires—3.50

Hoisting and hauling—5

Crane equipment—6

Freight elevator—8 to 10

Passenger elevator—10 to 12

The design procedure for cable and sheave selection is as follows:

Step 1
Determine the design load. This is a combination of:

The load to be lifted (live load)

Weight of the equipment (dead load)

Factor of safety

Thus

$$\text{Design load (lb)} = (\text{live load} + \text{dead load}) \times \text{factor of safety}$$

10-1

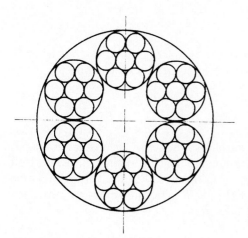

Fig. 10-1. Cable cross section.

Step 2

The minimum required cable area is determined by

$$S = P/A \qquad\qquad 10\text{-}2$$

Where S = breaking stress (psi)

P = design load (lb)

A = cable area (in.^{2})

The breaking stress is determined by the material used. As mild plow steel is the common type, 160,000 psi would be used.

Step 3

The area of a circular cross section is

$$A = d^2/4 = .7854\ d^2$$

This would be used to determine the minimum cable diameter if it were a solid strand. This is not the case, however, as illustrated in Fig. 10-1. Only the wires themselves are considered load-carrying sections. The remaining area helps the bending process. This accounts for nearly half the total area. Thus for design purpose use

$$A = .40\ dc^2 \qquad\qquad 10\text{-}3$$

Where A = minimum required area (in.^{2})

.40 = the portion of the section which carries load

dc = minimum cable diameter (in.)

Step 4

If the cable must carry a movable product, a sheave will be required. Its minimum size will be based upon the type and diameter of cable used. The three considered in this chapter are:

6 × 7 cable: minimum sheave ϕ = 42 dc

6 × 19 cable: minimum sheave ϕ = 30 dc

6 × 37 cable: minimum sheave ϕ = 18 dc

Step 5

Additional components such as shafts, keys, and bearings may be designed. Use the procedures developed in previous chapters.

It may not always be possible to determine the cable load by inspection. The induced force on the hoisting cable of Fig. 10-2 is easily determined (T lb). The minimum cable size may be found by solving Eqs. 10-1, 10-2, and 10-3. The force induced in the tie, however, is not easily seen. The principles of mechanic vectors must be applied. That is, the sum of the vertical forces must equal zero. There are 2T lb acting down, so there must be 2T lb counteracting this. Break the tie into horizontal and vertical com-

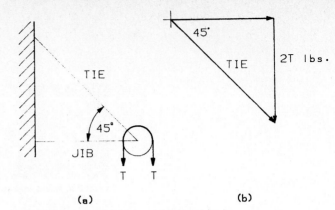

Fig. 10-2 (a and b). *Tie, jib, and hoist.*

ponents as shown in Fig. 10-2(b). Use right-angle trigonometry to find the tie force.

$$F = 2T/\text{sine }45° = 2T/.707\text{ lb}$$

The minimum required tie diameter may now be determined by solving Eqs. 10-1, 10-2, and 10-3.

DESIGN APPLICATION

EXAMPLE 1

SPECIFICATION

Design a cable system to lift a small passenger elevator. The total weight, live and dead load, is 2500 lb.

SOLUTION

Step 1

Since the application involves personal safety, use a 12:1 factor. The design load is 2500 × 12 = 30,000 lb.

Step 2

Determine the required cable area by the tensile Eq. 10-2:

$$S = P/A$$

If mild plow steel is used, the ultimate stress is 160,000 psi and the required area becomes

$$A = P/S = 30,000/160,000 = .187\text{ in.}^2$$

Step 3

The area of a circle is .7854 d^2. With cable, however, the area must be that of the load-carrying portion known as the *metallic area*. For the cables this area is approximately .40 d. The minimum required cable diameter becomes

$$dc = \sqrt{A/.4} = \sqrt{.187/.4} = .68\text{ in.}$$

Cable is available in .12-in. increments up to a diameter of 1.00 in. Use the next largest standard size, which is ϕ.75 in.

Step 4

If 6×19 cable is used, the minimum matching sheave diameter is

$$30 \times dc = 30 \times .75 = \phi 22.50 \text{ in.}$$

EXAMPLE 2

SPECIFICATION

The tension in the hoist cable shown in Fig. 10-2 is 1000 lb. Determine the minimum size of the tie.

SOLUTION

The force in the tie is

$$F = 2T/\text{sine } 45° = 2,000/.707 = 2828 \text{ lb}$$

The tie is used as a guy wire to hold the hoist in position. Thus the factor of safety is 3.50 and the design load is

$$2828 \times 3.50 = 9898 \text{ lb or approximately } 9900 \text{ lb}$$

The minimum required cable area is determined by

$$A = P/S = 9,900/160,000 = .06 \text{ in.}^2$$

$$dc = \sqrt{.06/.40} = \phi .40 \text{ in.}$$

Use a standard size ϕ.50-in. cable.

CADD DRAWING

Create the assembly for the Example 1 cable system. The drawing will not be to scale since the elevator size and elevation of lift are not known.

1. Boot the system, and SETUP C size drawing.
2. Select LINE, and create the elevator outline as shown in Fig. 10-3(a).
3. Create the cable, sheave, and attachments as shown in Fig. 10-3(b). Use LINE, CIRCLE, and CENTERLINE.
4. Before drawing the shafts and bearings, determine sizes by the simple beam reaction formula:

$$R1 = T/2 = 2500/2 = 1250 \text{ lb}$$

Select bearings from Appendix 7 that have an allowable load of at least this value using the least rpm column.
5. Create the top view using LINE, CENTERLINE, and SOLID CIRCLE.

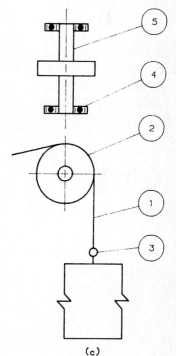

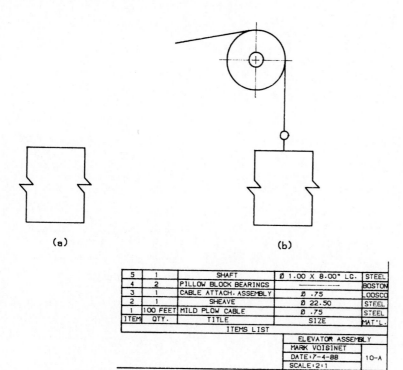

(a)   (b)   (c)

5	1	SHAFT	Ø 1.00 X 8.00" LG.	STEEL
4	2	PILLOW BLOCK BEARINGS		BOSTON
3	1	CABLE ATTACH. ASSEMBLY	Ø .75	LOOSCO
2	1	SHEAVE	Ø 22.50	STEEL
1	100 FEET	MILD PLOW CABLE	Ø .75	STEEL
ITEM	QTY.	TITLE	SIZE	MAT'L.
		ITEMS LIST		

ELEVATOR ASSEMBLY	
MARK VOISINET	
DATE:7-4-88	10-A
SCALE:2:1	

(d)

Fig. 10-3 (a–d). Elevator assembly.

6. Identify each item on the drawing with LEADER, NOTE, and CIRCLE. The result is shown in Fig. 10-3(c).

7. ZOOM in at the lower right portion of the drawing. Create the outline of a title block and an items list.

8. Select TEXT and .12 HEIGHT. Complete and fill in the title block and items list information as shown in Fig. 10-3(d).

ASSIGNMENTS

1. Design a cable and sheave to operate a hoist as follows:
   a. Provide maximum flexibility.
   b. Live load (material weight) is 20,000 lb.
   c. Dead load (container weight) is 3500 lb.
   Present the information on a small not-to-scale drawing.
2. a. Design a cable system to lift a freight elevator, total weight of 4000 lb. Use mild plow steel and 6 × 19 cable.
   b. Create an assembly drawing and items list similar to the procedure shown in Fig. 10-3.
3. a. Graphically determine the force acting on the tie shown in Fig. 10-4. Graphically estimate the angle of each member.

   b. Determine the required size of the mild plow steel cable for the 7850-lb. cable and the 8-ft tie. What standard size cables should be used? What size pulley should be used with a 6 × 19 cable?
   c. Create an assembly drawing and items list.
4. a. Graphically determine the force acting on the tie as shown in Fig. 10-5. Estimate the angle of each member.
   b. Determine the required size for the 1450-lb cable and the 36-ft tie. What standard-size cables would be used? What size pulley would be used with a 6 × 19 cable?
   c. Create an assembly drawing and items list.
5. a. Figure 10-6 illustrates a partial blast furnace process. Concentrating on the mechanical aspects of the furnace charging portion, design the skip car and cable system to charge the furnace. It is 75 ft from ground elevation at a 60° incline. The skip car (charging car) is 6 ft long by 4 ft wide by 3 ft high. It is filled with 20 percent limestone having a density of 162 lb/ft³, plus 80 percent coke and ore weighing 94 lb/ft³. Design the cable system, including the top sheave and bottom drum.
   b. Design shafts and bearings to support the sheave and drum. Use the shaft formula for transmission of pulley line shafts.
   c. Show your result in the form of a two-view assembly drawing.

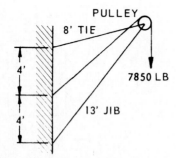

Fig. 10-4. Tie and jib.

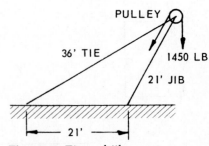

Fig. 10-5. Tie and jib.

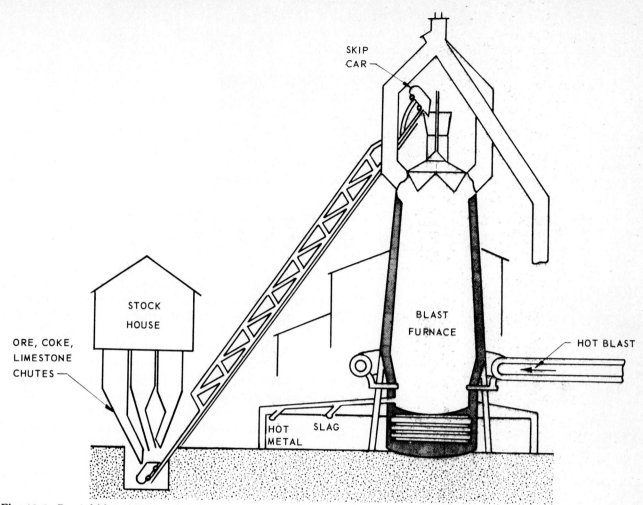

Fig. 10-6. *Partial blast furnace process.*

DESIGN PROJECTS

DRIVE SYSTEM SUMMARY

As you are now well aware, mechanical drive systems transmit:

- Power
- Motion

This is accomplished by one, or a combination of several, mechanical drive components. The standard components include:

- Belts
- Gears
- Compound gear trains
- Chains
- Cams
- Linkage
- Cable

In addition to the major components, other mechanical parts will be used in a machine. Each part must be individually sized to safely handle the specified power to be transmitted. These additional parts, found in virtually every machine, include:

- Shafts
- Keys
- Bearings
- Hardware (for example, retaining rings)

A chassis or housing will usually be formed around a machine. It will provide structural support, protect the parts, and maintain personal safety. A chassis, housing, or support will consist of a:

- Weldment
- Sheet-metal pattern
- Casting

A machine may be composed of any combination of parts listed in each category. So you can see there is virtually an unlimited variety. If you have studied each chapter diligently, however, you should feel secure enough to "tackle" such machine design problems. This chapter will help provide experience in "putting together" a machine design project. Think of it as designing a single component. The only difference is it will be done several times. You will not be overwhelmed by thinking about one small segment of the design at a time. "Compartmenting" components will make it much easier to handle. Simply apply the design theory presented in the previous chapters. In the end you may become quite an accomplished mechanical machine designer using CADD.

DRIVE SELECTION

The type of system to be used may not always be specified. What mechanical drive should be used? There are many factors to consider. Some of the more common considerations include:

Power

Chain drives are the most rugged. They readily transfer heavy, bulky material loads. There is little tension on the slack side of a chain. Thus bearing loads are lower. This means smaller bearings with less maintenance. For combined high-speed and high-horsepower applications, gears are best to use.

Motion

Gears and chain provide a uniform speed ratio. The output rpm is constant and it is easily determined. Gears are best for both high-speed and high-speed-ratio applications. A chain sprocket with a small number of teeth may result in a "jerky" motion. Belts and cable do not provide a constant speed ratio and are used when this is not a critical factor. Belt slippage can pose problems due to both a speed and power loss. Also, since the power is transmitted via friction to the pulley, a higher arc of contact (150° minimum) is necessary.

Center Distance

Gear center distance is restricted since it must conform to the pitch radii. They are best for compact applications. Chain, cable, and linkages have unrestricted centers and are best for large applications. Belt drives are somewhat restrictive falling within the intermediate range.

Cost

The cost of each type of system will generally be (in descending order) as follows:

Gears (most expensive)

Chain—roller

Cam

Belts—V

Cable

Linkage (least expensive)

Maintenance

Chains have loose tolerances and are easy to install. There are three maintenance methods. The oil-bath type will require a sheet-metal housing. Because of a rolling action and the fact that several teeth are engaged to help carry the load, wear is reduced. Gears require lubrication only. If constant lubrication is desired, a cast housing will be required. Belts do not require lubrication. Occasionally, however, the centers need to be adjusted. This is due to the belt stretching during use. They are inexpensive and easy to replace.

Service Factor

Belts are less noisy and have excellent shock absorption capability. Service, however, will be negatively affected when exposed to sun, oil, or other environmental conditions. Chains are not sensitive to these conditions. They are not temperature sensitive, and service is not affected by exposure to sunlight. Chain systems, however, can produce excessive noise.

It is not always possible to readily determine the type of drive system to use. As you can see, there are many factors affecting this decision. Only common ones were covered in this section. Should you have to make a decision of this nature, these factors will provide valuable assistance. Use them as a guide.

PROJECT CONCEPTS

An industrial design project may encompass any combination of the topics presented in this book, and more. The complete project concept may deal with subject matter beyond the scope presented herein. For example, to fully complete the power transmission projects, the electrical requirements would have to be investigated. This would include preparation of the power drawings involving much additional theory, including such topics as selection of the required conductor and conduit sizes. It may also include the preparation of the electrical control drawing package. Again, this is another complete set of theory applications and can be learned in additional technical courses. Different portions of design projects are normally handled by different departments within a design office. This book addresses that portion which would fall within the domain of the mechanical design department. It should be stressed, however, that it would be extremely advantageous for you to take technical courses in related disciplines. This will help you gain the perspective that will enable you to understand the complete project. Taking courses is especially important for those wishing to advance rapidly and those who work in small companies where versatility is important. It is an accomplishment, nonetheless, to apply mechanical design theory to a project.

PROCEDURE

Step 1
It is often necessary to decide beforehand which concepts must be applied. For example, as previously indicated, power and motion may be transmitted from one shaft to another by a variety of ways, namely, by belts, gears, chains, cams, linkages, cables, or a combination of these. Many factors in selecting the drive are considered, such as center distance, timing, cost, maintenance, power required, speed changes, installation, atmospheric conditions, noise, etc. Refer to the section "Drive Selection" for comparative data.

Step 2
Once the drive has been selected, the support components must be designed. These include the shafts, keys, couplings, bearings, seals, and other hardware either of a permanent (welding) or nonpermanent (fasteners/rings) nature. Remember to think about one design at a time.

Step 3

The enclosure is next considered. It may be a casting, sheet steel, structural steel, or a combination of more than one. Additional mechanics, strength of materials, and descriptive geometry concepts and application may be required.

The mechanical portion of a design and drafting project itself becomes quite involved and complex. The above format can even be expanded, since additional components may be required for use in systems such as piping and fluid power. The following assignments are presented to investigate and complete the mechanical part of a design drafting project.

PROJECT ASSIGNMENTS

1. Prepare a mechanical design using three different (gears, chain, belt) components for the system shown in Fig. 11-1. Use the following data:

	GEAR REDUCER (rpm)			
MOTOR	IN	OUT	MACH 1	MACH 2
10 hp at 1160 rpm	800	360 ± 5%	Design .50 hp at 110 rpm ± 3%	Design 5.00 hp at 180 ± 5%

  a. Determine which drive shall be the belt, which the gear, and which the chain. Size the components for each system.
  b. Select a gear reducer to meet the above requirements.
  c. Size all shafts and keys.
  d. Prepare a line assembly drawing. Include callouts for each component.

2. Determine the size of each of the bearings shown in Fig. 11-2. Select the best type of drive based upon minimum bearing loads. Shaft A must be capable of transmitting 15.00 hp at 1750 rpm. The final output rpm should be close to 72.
  a. Size the two-drive systems. A standard 4:1 speed reducer is used for the intermediate shafts.
  b. Determine the radial loads acting on each bearing.
  c. Size each bearing for an average life of 7500 h or more.
  d. Prepare a line assembly diagram. Include all design data on the assembly.

3. a. The drill press shown in Fig. 11-3 is powered by either of the following motors (as assigned by instructor):
    .50 hp at 1750 rpm;
    .38 hp at 1160 rpm.
    Design the pulley and belt drive to withstand system requirements. Use step pulleys for different speeds. Determine and specify allowable horsepower at each speed. Determine and specify the output speeds at the spindle.
  b. Prepare a head assembly drawing which includes the rack and spindle subassembly from assignment 6-3.
  c. Prepare an item list for part b. Select standard manufacturer parts for all purchased items.
  d. Determine the critical dimensions for a head casting from the assembly in part b. Refer to the casting section in Chapter 2.

4. a. The system shown in Fig. 11-4 is driven by a .50-hp, 1160-rpm motor mounted on an adjustable base. Use minimum center distances for the first and second drives with (approximately) a 24-in. center distance for the final drive. Step pulleys with at least three different diameter grooves each are used for the second drive. When the belt is located on the center-step pulley grooves, the driven machine will turn (as closely as possible) at 250 rpm.
    Design the three best drive systems. Each operates under normal duty. Determine and specify allowable horsepower. Determine exact speeds within the system and available at the driven machine.
  b. Determine total component forces and solve actual bearing loads. Select bearings from Appendix 7.
  c. Design shafts and match them to the bearing and pulley bores.
  d. Present calculations and rationale to support each aspect of the design.
  e. Prepare an assembly drawing. Include location dimensions and elevations as necessary.
  f. Prepare an items list using standard manufacturer components.

5. a. The following table provides required horsepower and speed specifications at each shaft as shown in Fig. 11-5. There are four sets of data. Complete the one assigned by the instructor.

	HP	RPM	HP	RPM	HP	RPM	HP	RPM
S1 motor	5	1750	7.50	3600	3	1160	3	1750
S2 gear reduction	5	1750	7.50	2100	3	800	3	1150
S3 gear reduction	5	?	7.50	?	3	?	3	?
S4 gear reduction	5	240	7.50	320	3	160	3	210
S5 machine A	2.50	48	3	110	2	110	1.50	92
S6 machine B	2	60	4	106	1	80	1.50	105

Design gears, shafts, keys, and bearings to withstand system requirements. Complete the tables shown in Fig. 11-5. Match the shafts to the gear and bearing bores.

b. Prepare a sectioned assembly drawing.

c. Prepare an items list. Include standard manufacturer parts for the gears, motor, bearings, couplings, seals, and gaskets.

d. Determine the critical dimensions required for the cast housing.

6. a. Select a motor and gear reducer combination to drive the two 6-ft gears shown in the cereal flaker Fig. 11-6. Drive the system with a standard 30-hp motor. Each drum shall rotate as close as possible to 29 rpm.

b. Design the gears to withstand the system requirements.

c. Design the shafts and bearings to withstand the system requirements.

d. Prepare an assembly drawing. Include two 8-ft-long drums with a 0.5-in. spacing between their outer surfaces.

e. Prepare a working drawing of a sheet-metal vapor hood which will extend over both drums.

7. The conveyor shown in Fig. 11-7(a) is used to transfer 4-in.- × 6-in.- × 20-ft-long steel billets at a surface speed of approximately 80 ft/min. The conveyor center-to-center distance is approximately 45 ft, so that it is possible to load it with two full billets. Steel weighs .28 lb/in.^{3} The conveyor operates 8 h/day.

a. Using a slat conveyor without sides, design the approximate chain and sprocket system. Double-strand chain (having a 1.7 strand factor) may be used as shown in Fig. 11-7(b). Use the following design process.

(i) Choose a tentative surface speed of 80 to 100 ft/min.

(ii) Determine total chain pull (weight of billets, approximate conveyor slat weight, and chain weight). Use .50 plate steel slats. Each chain and attachments will weigh approximately 6.00 lb/ft.

(iii) Determine the required motor horsepower (hp = (total weight × surface speed)/33,000).

(iv) Determine the required design horsepower if the system operates under moderate shock.

(v) Select sprockets and chain from the appendix tables. Begin your design using a trial rpm of 20.

(vi) Determine the actual drive sprocket rpm from the surface speed and circumference of the sprocket pitch diameter since the rpm will likely not be 20.

(vii) Redo steps 5 and 6 with the new rpm value. This is a trial-and-error method to match the sprocket size to the rpm.

b. From the horsepower and speed requirements, select a suitable motor and speed reducer combination.

c. Design the head and tail shaft plus the pillow block bearings for these shafts.

d. Select a standard chain takeup for adjustment. Provide chain roller supports at every 80 pitches. The bearings for these supports do not require design since the forces will be low.

e. Prepare an assembly drawing, showing all necessary components including slats, slat attachment to chain, and roller supports, similar to that shown in Fig. 11-7(c).

f. Prepare an items list for e using as many standard parts as possible.

g. Prepare detail drawings from the assembly drawing.

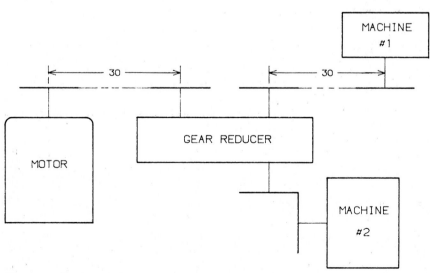

Fig. 11-1. Design project.

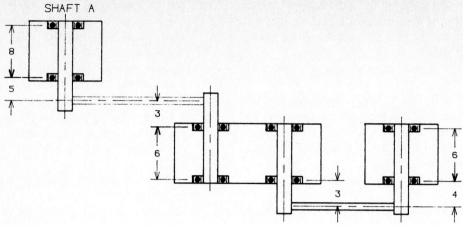

Fig. 11-2. Design project.

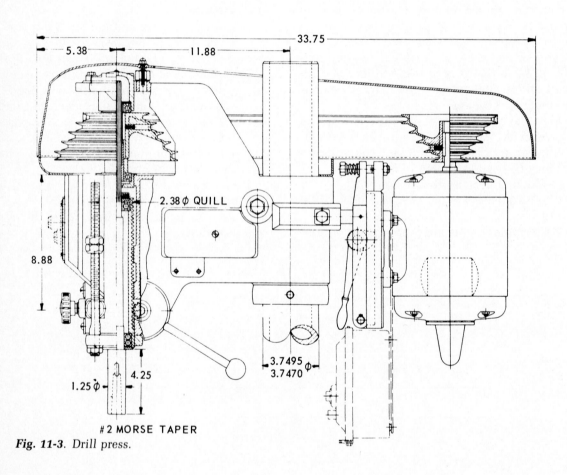

Fig. 11-3. Drill press.

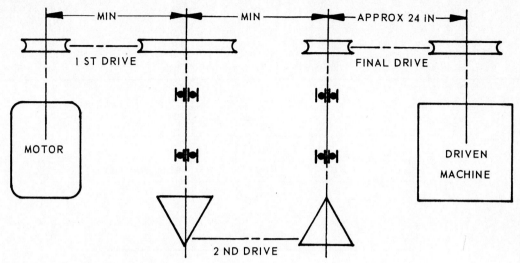

Fig. 11-4. *V-belt drive system.*

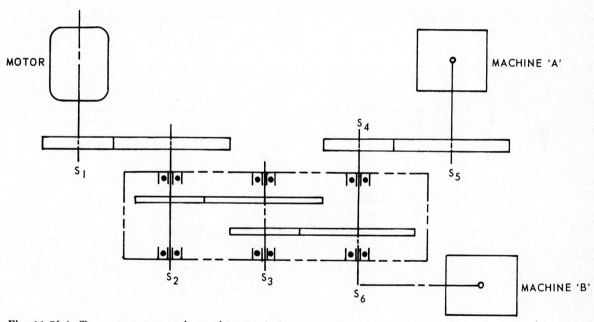

Fig. 11-5(a). *Ten gear power and speed transmission.*

TABLE 1 GEAR DATA

Gear	Speed Ratio	Number of Teeth	Pitch	Allowable HP	Bore	Hub Diameter	Hub Length
1							
2							
3							
•							
•							
9							
10							

TABLE 2 SHAFT DATA

Shaft	RPM	Reqd Dia	Allowable HP
1			
•			
•			
•			
6			

TABLE 3 KEY DATA

Key No	RPM	Sq Size	Reqd Length	Actual Length
1				
2				
•				
•				
9				
10				

TABLE 4 BEARING DATA

Bearing No	Bore	Actual Force	Allowable Force	Avg Life
1				
2				
•				
•				
5				
6				

Fig. 11-5(b). Tables.

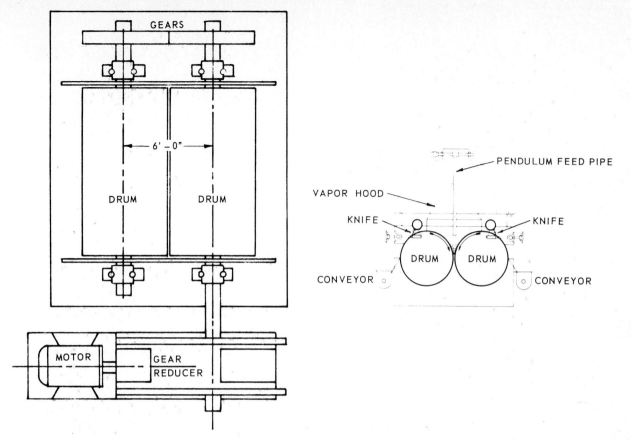

Fig. 11-6. Rotary drum dryer and flaker drive.

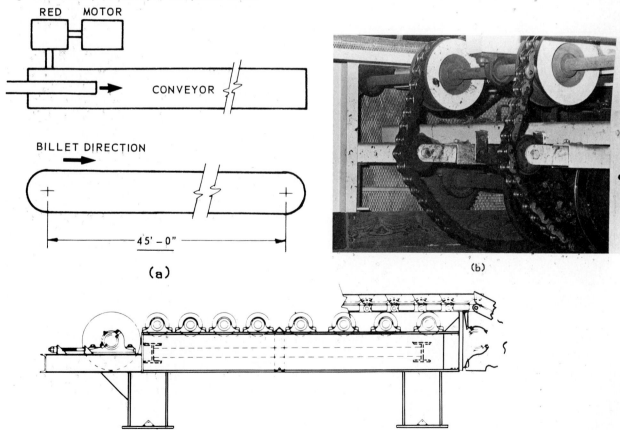

Fig. 11-7. (a) Horizontal slat conveyor; (b) double-strand chain; (c) detail of conveyor system.

APPENDICES

STANDARD NEMA* MOTORS
NEMA FRAME AND HP ASSIGNMENTS FOR
OPEN TYPE, POLYPHASE
SQUIRREL-CAGE 60 CYCLE MOTORS

HP	3600 RPM (3500) FRAME NO. T	1800 RPM (1750) FRAME NO. T	1200 RMP (1160) FRAME NO. T	900 RPM (875) FRAME NO. T
.50	. . . . .	. . . . .	. . . . .	143T
.75	. . . . .	. . . . .	143T	145T
1.00	. . . . .	143T	145T	182T
1.50	143T	145T	182T	184T
2	145T	145T	184T	213T
3	145T	182T	213T	215T
5	182T	184T	215T	254T
7.50	184T	213T	254T	256T
10	213T	215T	256T	284T
15	215T	254T	284T	286T
20	254T	256T	286T	324T
25	256T	284T	324T	326T
30	284TS	286T	326T	364T
40	286TS	324T	364T	365T
50	324TS	326T	365T	404T
60	326TS	364TS	404T	405T
75	364TS	365TS	405T	444T
100	365TS	404TS	444T	445T
125	404TS	405TS	445T	. . . . .
150	405TS	444TS	. . . . .	. . . . .
200	444TS	445TS	. . . . .	. . . . .
250	445TS	. . . . .	. . . . .	. . . . .

*NEMA—National Electrical Manufacturers Association

Single Groove Sheaves
COMBINATION GROOVE
for "4L" or "A" Belts and "5L" or "B" Belts
Stock Sizes—Finished Bore

21/32"

Table No. 1

Part No.	Outside	Pitch "A"	Pitch "B"	Type	1/2"	5/8"	3/4"	7/8"	15/16"	1"	1 1/8"	1 3/16"	1 1/4"	1 3/8"	1 7/16"	F	L	P	C	Wt. Lbs.
▲BK24	2.40"	1.80"	2.20"	1	X	X	X	—	—	—	—	—	—	—	—	13/16"	1 1/16"	13/32"	5/32"	.4
▲BK25	2.50	1.90	2.30	1	X	X	X	X	—	—	—	—	—	—	—	13/16	1 1/16	13/32	5/32	.5
▲BK26	2.60	2.00	2.40	1	X	X	X	X	—	—	—	—	—	—	—	13/16	1 1/16	13/32	5/32	.6
▲BK27	2.70	2.10	2.50	2	X	X	X	X	—	—	—	—	—	—	—	13/16	1 1/16	13/32	5/32	.6
▲BK28	2.95	2.20	2.60	2	X	X	X	X	—	—	—	—	—	—	—	13/16	1 1/16	13/32	5/32	.8
▲BK30	3.15	2.40	2.80	2	X	X	X	X	—	—	—	—	—	—	—	13/16	1 1/16	13/32	5/32	.8
▲BK32	3.35	2.60	3.00	2	X	X	X	X	—	—	—	—	—	—	—	13/16	1 1/16	13/32	5/32	.8
BK34	3.55	2.80	3.20	2	X	X	X	X	—	X	X	—	—	—	—	7/8	15/32	13/32	1/8	1.3
BK36	3.75	3.00	3.40	2	X	X	X	X	—	X	X	—	—	—	—	7/8	15/32	13/32	1/8	1.5
BK40	3.95	3.20	3.60	2	X	X	X	X	—	X	X	—	—	—	—	7/8	15/32	13/32	1/8	1.5
BK45	4.25	3.50	3.90	2	X	X	X	X	—	X	X	—	—	—	—	7/8	15/32	13/32	1/8	1.8
BK47	4.45	3.70	4.10	2	X	X	X	X	—	X	X	—	—	—	—	7/8	15/32	13/32	1/8	1.9
BK50	4.75	4.00	4.40	3	X	X	X	X	X	X	X	—	—	—	—	7/8	15/32	13/32	1/8	2.0
BK52	4.95	4.20	4.60	3	X	X	X	X	—	X	X	—	—	—	—	7/8	15/32	13/32	1/8	2.0
BK55	5.25	4.50	4.90	3	X	X	X	X	—	X	X	X	—	—	—	7/8	15/32	13/32	1/8	2.2
BK57	5.45	4.70	5.10	3	—	X	X	X	X	X	X	—	—	—	—	7/8	15/32	13/32	1/8	2.3
BK60	5.75	5.00	5.40	3	X	X	X	X	—	X	X	X	—	—	—	7/8	15/32	13/32	1/8	2.3
BK62	5.95	5.20	5.60	3	X	X	X	X	X	X	X	X	—	—	—	7/8	15/32	13/32	1/8	2.4
BK65	6.25	5.50	5.90	3	—	X	X	—	—	X	X	—	—	—	—	7/8	15/32	13/32	1/8	2.7
BK67	6.45	5.70	6.10	3	—	X	X	—	—	X	X	—	—	—	—	7/8	15/32	13/32	1/8	2.8
BK70	6.75	6.00	6.40	3	—	X	X	—	X	X	X	—	X	—	X	7/8	1 15/32	21/32★	1/16★	3.3
BK72	6.95	6.20	6.60	3	—	—	—	—	—	X	X	—	—	X	—	7/8	1 15/32	21/32	1/16	3.9
BK75	7.25	6.50	6.90	3	—	—	—	—	—	X	X	—	—	X	—	7/8	1 15/32	21/32	1/16	3.9
BK77	7.45	6.70	7.10	3	—	—	—	—	—	X	X	—	—	X	—	7/8	1 15/32	21/32	1/16	4.1
BK80	7.75	7.00	7.40	3	—	—	X	X	X	—	X	X	X	X	X	7/8	1 15/32	21/32	1/16	4.4
BK85	8.25	7.50	7.90	3	—	—	X	—	X	—	X	—	—	X	X	7/8	1 15/32	21/32	1/16	5.0
BK90	8.75	8.00	8.40	3	—	—	X	X	X	X	X	X	—	X	X	7/8	1 15/32	21/32	1/16	5.0
BK95	9.25	8.50	8.90	3	—	—	X	—	X	X	X	—	—	X	X	7/8	1 15/32	21/32	1/16	5.4
BK100	9.75	9.00	9.40	3	—	—	X	X	X	X	X	X	X	X	X	7/8	1 15/32	21/32	1/16	5.6
BK105	10.25	9.50	9.90	3	—	—	—	—	X	X	X	—	X	X	X	7/8	1 15/32	21/32	1/16	5.8
BK110	10.75	10.00	10.40	3	—	—	X	—	X	X	X	—	X	X	X	7/8	1 15/32	21/32	1/16	6.4
BK115	11.25	10.50	10.90	3	—	—	—	—	—	X	—	—	X	X	X	7/8	1 15/32	21/32	1/16	6.9
BK120	11.75	11.00	11.40	3	—	—	X	—	X	X	—	X	—	X	X	7/8	1 15/32	21/32	1/16	7.4
BK130	12.75	12.00	12.40	3	—	—	X	—	X	X	X	X	—	X	X	7/8	1 15/32	21/32	1/16	8.4
BK140	13.75	13.00	13.40	3	—	—	X	—	X	X	—	X	—	X	X	7/8	1 15/32	21/32	1/16	9.4
BK160	15.75	15.00	15.40	3	—	—	—	—	—	X	X	X	X	—	X	7/8	1 15/32	21/32	1/16	11.4
BK190	18.75	18.00	18.40	3	—	—	—	—	—	X	—	X	X	—	X	7/8	1 15/32	21/32	1/16	13.4

Other Bores can be furnished in production quantities only. Prices on application.

★P = 13/32" and C = 1/8" for 1" Bores and Smaller.

▲NOTE—DO NOT USE THESE "BK" SHEAVES WITH "B" GRIPNOTCH BELT RATINGS.

These sheaves are also available as "BG" Sheaves; order by Part Number "BG24 x 1/2", etc.

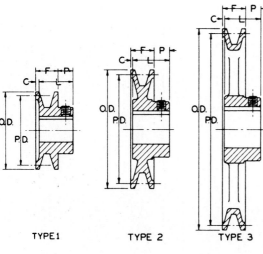

TYPE 1   TYPE 2   TYPE 3

♦ ♦

HOLLOW HEAD SETSCREWS

♦ ♦

STATICALLY BALANCED

♦ ♦

Standard Keyseats
Table No. 2

Bore Range	Keyseat
1/2"	None
5/8 to 7/8	3/16" x 3/32"
15/16 to 1 1/4	1/4 x 1/8
1 7/16	3/8 x 3/16

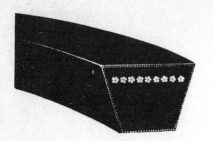

"A" Belts
½″ x ⁵⁄₁₆″

"B" Belts
²¹⁄₃₂″ x ⁷⁄₁₆″

"C" Belts
⁷⁄₈″ x ¹⁷⁄₃₂″

"D" Belts
1 ¼″ x ¾″

Table No. 1

Stock Sizes

Belt No.	Outside	Pitch	Wt. Lbs.	Belt No.	Outside	Pitch	Wt. Lbs.	Belt No.	Outside	Pitch	Wt. Lbs.	Belt No.	Outside	Pitch	Wt. Lbs.
A23	25.2″	24.3″	.2	A89	91.2″	90.3″	.6	B70	73.0″	71.8″	.8	C60	64.2″	62.9″	1.2
A24	26.2	25.3	.2	A90	92.2	91.3	.6	B71	74.0	72.8	.8	C68	72.2	70.9	1.3
A25	27.2	26.3	.2	A91	93.2	92.3	.6	B72	75.0	73.8	.8	C72	76.2	74.9	1.4
A26	28.2	27.3	.2	A92	94.2	93.3	.6	B73	76.0	74.8	.8	C75	79.2	77.9	1.4
A27	29.2	28.3	.2	A93	95.2	94.3	.6	B74	77.0	75.8	.8	C81	85.2	83.9	1.6
A28	30.2	29.3	.2	A94	96.2	95.3	.6	B75	78.0	76.8	.8	C85	89.2	87.9	1.6
A29	31.2	30.3	.2	A95	97.2	96.3	.6	B76	79.0	77.8	.8	C90	94.2	92.9	1.7
A30	32.2	31.3	.2	A96	98.2	97.3	.7	B77	80.0	77.8	.8	C96	100.2	98.9	1.8
A31	33.2	32.3	.2	A97	99.2	98.3	.7	B78	81.0	79.8	.8	C97	101.2	99.9	1.8
A32	34.2	33.3	.2	A98	100.2	99.3	.7	B79	82.0	80.8	.8	C99	103.2	101.9	1.9
A33	35.2	34.3	.2	A100	102.2	101.3	.7	B80	83.0	81.8	.9	C100	104.2	102.9	1.9
A34	36.2	35.3	.2	A103	105.2	104.3	.7	B81	84.0	82.8	.9	C105	109.2	107.9	2.0
A35	37.2	36.3	.2	A105	107.2	106.3	.7	B82	85.0	83.8	.9	C108	112.2	110.9	2.0
A36	38.2	37.3	.3	A110	112.2	111.3	.8	B83	86.0	84.8	.9	C109	113.2	111.9	2.0
A37	39.2	38.3	.3	A112	114.2	113.3	.8	B84	87.0	85.8	.9	C112	116.2	114.9	2.1
A38	40.2	39.3	.3	A120	122.2	121.3	.8	B85	88.0	86.8	.9	C115	119.2	117.9	2.1
A39	41.2	40.3	.3	A128	130.2	129.3	.9	B86	89.0	87.8	1.0	C120	124.2	122.9	2.3
A40	42.2	41.3	.3	A136	138.2	137.3	.9	B87	90.0	88.8	1.0	C124	128.2	126.9	2.4
A41	43.2	42.3	.3	A144	146.2	145.3	1.0	B88	91.0	89.8	1.0	C128	132.2	130.9	2.4
A42	44.2	43.3	.3	A158	160.2	159.3	1.1	B89	92.0	90.8	1.0	C136	140.2	138.9	2.6
A43	45.2	44.3	.3	A173	175.2	174.3	1.2	B90	93.0	91.8	1.0	C144	148.2	146.9	2.8
A44	46.2	45.3	.3	A180	182.2	181.3	1.2	B91	94.0	92.8	1.0	C150	154.2	152.9	2.9
A45	47.2	46.3	.3	B25	28.0	26.8	.3	B92	95.0	93.8	1.0	C158	162.2	160.9	3.0
A46	48.2	47.3	.3	B26	29.0	27.8	.3	B93	96.0	94.8	1.0	C162	166.2	164.9	3.1
A47	49.2	48.3	.3	B28	31.0	29.8	.3	B94	97.0	95.8	1.0	C173	177.2	175.9	3.3
A48	50.2	49.3	.3	B29	32.0	30.8	.3	B95	98.0	96.8	1.0	C180	184.2	182.9	3.4
A49	51.2	50.3	.4	B30	33.0	31.8	.3	B96	99.0	97.8	1.1	C195	199.2	197.9	3.7
A50	52.2	51.3	.4	B31	34.0	32.8	.3	B97	100.0	98.8	1.1	C210	214.2	212.9	4.0
A51	53.2	52.3	.4	B32	35.0	33.8	.3	B98	101.0	99.8	1.1	C225	227.2	225.9	4.3
A52	54.2	53.3	.4	B33	36.0	34.8	.4	B99	102.0	100.8	1.1	C240	242.2	240.9	4.6
A53	55.2	54.3	.4	B34	37.0	35.8	.4	B100	103.0	101.8	1.1	C255	257.2	255.9	4.9
A54	56.2	55.3	.4	B35	38.0	36.8	.4	B101	104.0	102.8	1.1	C270	272.2	270.9	5.2
A55	57.2	56.3	.4	B36	39.0	37.8	.4	B103	106.0	104.8	1.1	C285	287.2	285.9	5.4
A56	58.2	57.3	.4	B37	40.0	38.8	.4	B105	108.0	106.8	1.1	C300	302.2	300.9	5.7
A57	59.2	58.3	.4	B38	41.0	39.8	.4	B108	111.0	109.8	1.2	C315	317.2	315.9	6.0
A58	60.2	59.3	.4	B39	42.0	40.8	.4	B111	114.0	112.8	1.2	C330	332.2	330.9	6.3
A59	61.2	60.3	.4	B40	43.0	41.8	.5	B112	115.0	113.8	1.2	C345	347.2	345.9	6.6
A60	62.2	61.3	.4	B41	44.0	42.8	.5	B116	119.0	117.8	1.3	C360	362.2	360.9	6.9
A61	63.2	62.3	.4	B42	45.0	43.8	.5	B120	123.0	121.8	1.3	C390	392.2	390.9	7.5
A62	64.2	63.3	.4	B43	46.0	44.8	.5	B124	127.0	125.8	1.3	C420	422.2	420.9	8.0
A63	65.2	64.3	.4	B44	47.0	45.8	.5	B128	131.0	129.8	1.4	D120	125.2	123.3	4.0
A64	66.2	65.3	.4	B45	48.0	46.8	.5	B133	136.0	134.8	1.5	D128	133.2	131.3	4.4
A65	67.2	66.3	.5	B46	49.0	47.8	.5	B136	139.0	137.8	1.5	D144	149.2	147.3	5.0
A66	68.2	67.3	.5	B47	50.0	48.8	.5	B140	143.0	141.8	1.6	D158	163.2	161.3	5.3
A67	69.2	68.3	.5	B48	51.0	49.8	.5	B144	147.0	145.8	1.6	D162	167.2	165.3	5.5
A68	70.2	69.3	.5	B49	52.0	50.8	.6	B148	151.0	149.3	1.6	D173	178.2	176.3	5.8
A69	71.2	70.3	.5	B50	53.0	51.8	.6	B150	153.0	151.8	1.6	D180	185.2	183.3	6.0
A70	72.2	71.3	.5	B51	54.0	52.8	.6	B154	157.0	155.8	1.7	D195	200.2	198.3	6.3
A71	73.2	72.3	.5	B52	55.0	53.8	.6	B158	161.0	159.8	1.7	D210	215.2	213.3	6.8
A72	74.2	73.3	.5	B53	56.0	54.8	.6	B162	165.0	163.8	1.7	D225	227.7	225.8	7.1
A73	75.2	74.3	.5	B54	57.0	55.8	.6	B173	176.0	174.8	1.9	D240	242.7	240.8	7.7
A74	76.2	75.3	.5	B55	58.0	56.8	.6	B180	183.0	181.8	1.9	D255	257.7	255.8	8.1
A75	77.2	76.3	.5	B56	59.0	57.8	.6	B190	193.0	191.8	2.0	D270	272.7	270.8	8.9
A76	78.2	77.3	.5	B57	60.0	58.8	.7	B195	198.0	196.8	2.0	D285	287.7	285.8	9.8
A77	79.2	78.3	.5	B58	61.0	59.8	.7	B205	208.0	206.9	2.2	D300	302.7	300.8	10.5
A78	80.2	79.3	.5	B59	62.0	60.8	.7	B210	213.0	211.8	2.3	D315	317.7	315.8	10.2
A79	81.2	80.3	.5	B60	63.0	61.8	.7	B225	226.5	225.3	2.5	D330	332.7	330.8	10.6
A80	82.2	81.3	.5	B61	64.0	62.8	.7	B240	241.5	240.3	2.6	D345	347.7	345.8	11.0
A81	83.2	82.3	.5	B62	65.0	63.8	.7	B255	256.5	255.3	2.8	D360	362.7	360.8	11.5
A82	84.2	83.3	.6	B63	66.0	64.8	.7	B270	271.5	270.3	2.9	D390	392.7	390.8	12.4
A83	85.2	84.3	.6	B64	67.0	65.8	.7	B285	286.5	285.3	3.1	D420	422.7	420.8	13.4
A84	86.2	85.3	.6	B65	68.0	66.8	.7	B300	301.5	300.3	3.2	D480	482.7	480.8	15.8
A85	87.2	86.3	.6	B66	69.0	67.8	.7	B315	316.5	315.3	3.4	D540	542.7	540.8	17.2
A86	88.2	87.3	.6	B67	70.0	68.8	.7	B360	361.5	360.3	4.0	D600	602.7	600.8	19.1
A87	89.2	88.3	.6	B68	71.0	69.8	.7	C51	55.2	53.9	1.0	D660	662.7	660.8	22.0
A88	90.2	89.3	.6	B69	72.0	70.8	.8	C55	59.2	57.9	1.1				

U.S. CUSTOMARY (INCHES)						METRIC (MILLIMETERS)					
Diameter of Shaft		Square Key		Flat Key		Diameter of Shaft		Square Key		Flat Key	
		Nominal Size		Nominal Size				Nominal Size		Nominal Size	
From	To	W	H	W	H	Over	Up To	W	H	W	H
.500	.562	.125	.125	.125	.094	6	8	2	2		
.625	.875	.188	.188	.188	.125	8	10	3	3		
.938	1.250	.250	.250	.250	.188	10	12	4	4		
						12	17	5	5		
1.312	1.375	.312	.312	.312	.250	17	22	6	6		
1.438	1.750	.375	.375	.375	.250	22	30	7	7	8	7
1.812	2.250	.500	.500	.500	.375	30	38	8	8	10	8
						38	44	9	9	12	8
						44	50	10	10	14	9
						50	58	12	12	16	10

Square and flat stock keys.

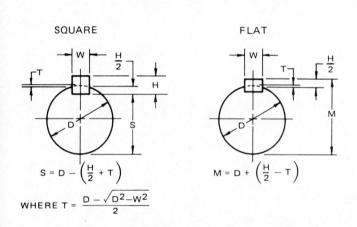

SQUARE

FLAT

$$S = D - \left(\frac{H}{2} + T\right)$$

$$M = D + \left(\frac{H}{2} - T\right)$$

$$\text{WHERE } T = \frac{D - \sqrt{D^2 - W^2}}{2}$$

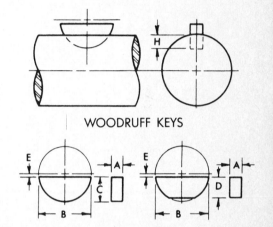

WOODRUFF KEYS

U.S. CUSTOMARY (INCHES)					METRIC (MILLIMETERS)					
Nominal Size	Key			Keyseat	Key No.	Nominal Size	Key			Keyseat
A × B	E	C	D	H		A × B	E	C	D	H
.062 × .500	.047	.203	.194	.172	204	1.6 × 12.7	1.5	5.1	4.8	4.2
.094 × .500	.047	.203	.194	.156	304	2.4 × 12.7	1.3	5.1	4.8	3.8
.094 × .625	.062	.250	.240	.203	305	2.4 × 15.9	1.5	6.4	6.1	5.1
.125 × .500	.049	.203	.194	.141	404	3.2 × 12.7	1.3	5.1	4.8	3.6
.125 × .625	.062	.250	.240	.188	405	3.2 × 15.9	1.5	6.4	6.1	4.6
.125 × .750	.062	.313	.303	.251	406	3.2 × 19.1	1.5	7.9	7.6	6.4
.156 × .625	.062	.250	.240	.172	505	4.0 × 15.9	1.5	6.4	6.1	4.3
.156 × .750	.062	.313	.303	.235	506	4.0 × 19.1	1.5	7.9	7.6	5.8
.156 × .875	.062	.375	.365	.297	507	4.0 × 22.2	1.5	9.7	9.1	7.4
.188 × .750	.062	.313	.303	.219	606	4.8 × 19.1	1.5	7.9	7.6	5.3
.188 × .875	.062	.375	.365	.281	607	4.8 × 22.2	1.5	9.7	9.1	7.1
.188 × 1.000	.062	.438	.428	.344	608	4.8 × 25.4	1.5	11.2	10.9	8.6
.188 × 1.125	.078	.484	.475	.390	609	4.8 × 28.6	2.0	12.2	11.9	9.9
.250 × .875	.062	.375	.365	.250	807	6.4 × 22.2	1.5	9.7	9.1	6.4
.250 × 1.000	.062	.438	.428	.313	808	6.4 × 25.4	1.5	11.2	10.9	7.9

NOTE: METRIC KEY SIZES WERE NOT AVAILABLE AT THE TIME OF PUBLICATION. SIZES SHOWN ARE INCH-DESIGNED KEY-SIZES SOFT CONVERTED TO MILLIMETERS. CONVERSION WAS NECESSARY TO ALLOW THE STUDENT TO COMPARE KEYS WITH SLOT SIZES GIVEN IN MILLIMETERS.

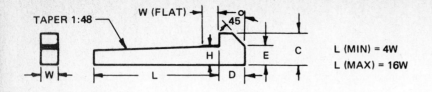

L (MIN) = 4W
L (MAX) = 16W

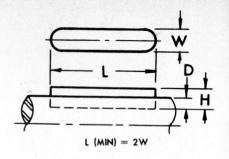

L (MIN) = 2W

U.S. CUSTOMARY (INCHES)

Shaft Diameter	Square Type					Flat Type				
	W	H	C	D	E	W	H	C	D	E
.500–.562	.125	.125	.250	.219	.156	.125	.094	.188	.125	.125
.625–.875	.188	.188	.312	.281	.219	.188	.125	.250	.188	.156
.938–1.250	.250	.250	.438	.344	.344	.250	.188	.312	.250	.188
1.312–1.375	.312	.312	.562	.406	.406	.312	.250	.375	.312	.250
1.438–1.750	.375	.375	.688	.469	.469	.375	.250	.438	.375	.312
1.812–2.250	.500	.500	.875	.594	.625	.500	.375	.625	.500	.438
2.312–2.750	.625	.625	1.062	.719	.750	.625	.438	.750	.625	.500
2.875–3.250	.750	.750	1.250	.875	.875	.750	.500	.875	.750	.625

METRIC (MILLIMETERS)

Shaft Diameter	Square Type					Flat Type				
	W	H	C	D	E	W	H	C	D	E
12–14	3.2	3.2	6.4	5.4	4	3.2	2.4	5	3.2	3.2
16–22	4.8	4.8	10	7	5.4	4.8	3.2	6.4	5	4
24–32	6.4	6.4	11	8.6	8.6	6.4	5	8	6.4	5
34–35	8	8	14	10	10	8	6.4	10	8	6.4
36–44	10	10	18	12	12	10	6.4	11	10	8
46–58	13	13	22	15	16	13	10	16	13	11
60–70	16	16	27	19	20	16	11	20	16	13
72–82	20	20	32	22	22	20	13	22	20	16

Note: Metric standards governing key sizes were not available at the time of publication. The sizes given in the above chart are "soft conversion" from current standards and are not representative of the precise metric key sizes which may be available in the future. Metric sizes are given only to allow the student to complete the drawing assignment.

Square and flat gib-head keys.

U.S. CUSTOMARY (INCHES)

Key No.	L	W	H	D
2	.500	.094	.141	.094
4	.625	.094	.141	.094
6	.625	.156	.234	.156
8	.750	.156	.234	.156
10	.875	.156	.234	.156
12	.875	.234	.328	.219
14	1.00	.234	.328	.234
16	1.125	.188	.281	.188
18	1.125	.250	.375	.250
20	1.250	.219	.328	.219
22	1.375	.250	.375	.250
24	1.50	.250	.375	.250
26	2.00	.188	.281	.188
28	2.00	.312	.469	.312
30	3.00	.375	.562	.375
32	3.00	.500	.750	.500
34	3.00	.625	.938	.625

METRIC (MILLIMETERS)

Key No.	L	W	H	D
2	12	2.4	3.6	2.4
4	16	2.4	3.6	2.4
6	16	4	6	4
8	20	4	6	4
10	22	4	6	4
12	22	6	8.4	7
14	25	6	8.4	6
16	28	5	7	5
18	28	6.4	10	6.4
20	32	7	8	5
22	35	6.4	10	6.4
24	38	6.4	10	6.4
26	50	5	7	5
28	50	8	12	8
30	75	10	14	10
32	75	12	20	12
34	75	16	24	16

Pratt and Whitney keys.

Standard Keys

STANDARD COUPLINGS

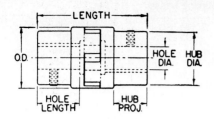

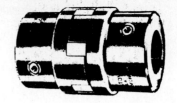

Shaft Dia. Range	Maximum Horsepower Rating at RPM of:†							Max. In. Lbs. Torque
	50	100	300	690	870	1150	1750	
$\frac{3}{8}$-$\frac{1}{2}$"	.16	.32	.95	2.2	2.8	3.6	5.6	200
$\frac{1}{2}$-$\frac{3}{4}$	.40	.79	2.4	5.5	6.9	9.1	13.9	500
$\frac{1}{2}$-1	.79	1.6	4.8	10.9	13.8	18.2	—	1000
$\frac{3}{4}$-1$\frac{1}{4}$	1.4	2.9	8.6	19.7	24.8	—	—	1800
1 -1$\frac{1}{2}$	2.5	5.1	15.2	35.0	—	—	—	3200
1$\frac{1}{4}$-1$\frac{7}{8}$	5.6	11.1	33.3	—	—	—	—	7000
1$\frac{3}{4}$-2$\frac{1}{8}$	8.7	17.5	—	—	—	—	—	11000

Size	Hole Lgth.	O.D.	Lgth.+	Hub Dia.	Hub Proj.	Keyway
20	1 7/16	2	3 11/16	1 3/4	1 1/8	1/8 x 1/16
20	1 7/16	2	3 11/16	1 3/4	1 1/8	3/16 x 3/32
20	1 7/16	2	3 11/16	1 3/4	1 1/8	3/16 x 3/32
20	1 7/16	2	3 11/16	1 3/4	1 1/8	3/16 x 3/32
20	1 7/16	2	3 11/16	1 3/4	1 1/8	1/4 x 1/8
20	1 7/16	2	3 11/16	1 3/4	1 1/8	1/4 x 1/8
25	1 19/32	2 1/2	4 1/8	2 1/4	1 1/4	3/16x3/32
25	1 19/32	2 1/2	4 1/8	2 1/4	1 1/4	3/16x3/32
25	1 19/32	2 1/2	4 1/8	2 1/4	1 1/4	1/4 x 1/8
25	1 19/32	2 1/2	4 1/8	2 1/4	1 1/4	1/4 x 1/8
25	1 19/32	2 1/2	4 1/8	2 1/4	1 1/4	1/4 x 1/8
25	1 19/32	2 1/2	4 1/8	2 1/4	1 1/4	1/4 x 1/8
30	2 5/32	3	5 15/32	2 3/4	1 11/16	1/4 x 1/8
30	2 5/32	3	5 15/32	2 3/4	1 11/16	1/4 x 1/8
30	2 5/32	3	5 15/32	2 3/4	1 11/16	1/4 x 1/8
30	2 5/32	3	5 15/32	2 3/4	1 11/16	5/16x5/32
30	2 5/32	3	5 15/32	2 3/4	1 11/16	3/8 x 3/16
30	2 5/32	3	5 15/32	2 3/4	1 11/16	3/8 x 3/16
38	2 5/8	3 3/4	6 5/16	3 1/2	1 7/8	1/4 x 1/8
38	2 5/8	3 3/4	6 5/16	3 1/2	1 7/8	3/8 x 3/16
38	2 5/8	3 3/4	6 5/16	3 1/2	1 7/8	3/8 x 3/16
38	2 5/8	3 3/4	6 5/16	3 1/2	1 7/8	3/8 x 3/16
38	2 5/8	3 3/4	6 5/16	3 1/2	1 7/8	3/8 x 3/16
38	2 5/8	3 3/4	6 5/16	3 1/2	1 7/8	1/2 x 1/4
45	3	4 1/2	7 3/16	4	2 1/8	3/8 x 3/16
45	3	4 1/2	7 3/16	4	2 1/8	1/2 x 1/4
45	3	4 1/2	7 3/16	4	2 1/8	1/2 x 1/4
45	3	4 1/2	7 3/16	4	2 1/8	1/2 x 1/4

(a)

ALL DIMENSIONS IN INCHES
ORDER BY CATALOG NUMBER OR ITEM CODE
To order complete coupling order two coupling halves

Coupling Size	Bore	Bore Length+	O.D.	Over-all Length++	Hub Dia.	Hub Proj.	Assembly Clearance†	Coupling Halves Catalog Number	Item Code	Coupling Inserts BOST-BRONZ Catalog Number	Item Code	
FC12	— 3/8 7/16 1/2	— 27/32	1-1/4	2-5/16	1	5/8	3-3/16	FC12 SOLID FC12—3/8 FC12—7/16 FC12—1/2	47448 08246 08248 08250	XFCBB12	08064	
FC15	— 1/2 9/16 5/8 3/4 7/8	— 1-1/32	1-1/2	2-3/4	1-1/4	3/4	3-3/4	FC15 SOLID FC15—1/2 FC15—9/16 FC15—5/8 FC15—3/4 FC15—7/8	47449 08252 61421 08254 08256 61422	XFCBB15	08066	
FC20	· 1/2 9/16 5/8 3/4 7/8 15/16 1	1-7/16	2	3-11/16	1-3/4	1-1/8	4-13/16	FC20 SOLID FC20—1/2 FC20—9/16 FC20—5/8 FC20—3/4 FC20—7/8 FC20—15/16 FC20—1	47450 08258 66063 08260 08262 08264 08266 08268	XFCBB20	08068	
FC25	— 3/4 7/8 1 1-1/8 1-3/16 1-1/4	— 1-19/32	2-1/2	4-1/8	2-1/4	1-1/4	5-3/8	FC25 SOLID FC25—3/4 FC25—7/8 FC25—1 FC25—1-1/8 FC25—1-3/16 FC25—1-1/4	47451 08270 08272 08274 08276 08278 08280	XFCBB25	08070	
FC30	1 1-1/8 1-1/4 1-3/8 1-7/16 1-1/2	2-5/32	3	5-15/32	2-3/4	1-11/16	7	FC30 SOLID FC30—1 FC30—1-1/8 FC30—1-1/4 FC30—1-3/8 FC30—1-7/16 FC30—1-1/2	47452 08282 08284 08286 08288 08290 08292	XFCBB30	08072	
FC38	— 1-1/4 1-1/2 1-9/16 1-5/8 1-3/4 1-7/8	— 2-5/8	3-3/4	6-5/16	3-1/2	1-7/8	8-3/16	FC38 SOLID FC38—1-1/4 FC38—1-1/2 FC38—1-9/16 FC38—1-5/8 FC38—1-3/4 FC38—1-7/8	24650 08294 08296 08298 08300 08302 08304	XFCBB38	08074	
FC45	1-3/4 1-7/8 2 2-1/8	3	4-1/2	7-3/16	4	2-1/8	9-5/16	FC45 SOLID FC45—1-3/4 FC45—1-7/8 FC45—2 FC45—2-1/8	24816 08306 08308 08310 08312	XFCBB45	08076	

+Length of hole in each half.
++Total length of coupling with jaws engaged full depth.
†Total length of coupling with jaws completely disengaged for insert assembly.

COLD FINISHED ROUNDS

LOW CARBON—CASE HARDENING

SIZE IN INCHES	WEIGHT PER FT. IN LBS.	SIZE IN INCHES	WEIGHT PER FT. IN LBS.
.12	.042	(Continued)	
.16	.065	2.75	20.19
.18	.094	2.81	21.12
.25	.167	2.88	22.07
.31	.261	2.94	23.04
.33	.288	3.00	24.03
.34	.316	3.06	25.05
.38	.376	3.12	26.08
.41	.441	3.18	27.13
.44	.511	3.25	28.21
.50	.668	3.38	30.42
.52	.710	3.44	31.55
.56	.845	3.50	32.71
.62	1.04	3.56	33.89
.69	1.26	3.62	35.09
.75	1.50	3.69	36.31
.77	1.56	3.75	37.55
.81	1.76	3.88	40.10
.88	2.04	3.99	41.40
.94	2.35	4.00	42.73
1.00	2.67	4.12	45.44
1.06	3.01	4.18	46.83
1.12	3.38	4.25	48.23
1.18	3.77	4.38	51.11
1.25	4.17	4.44	52.58
1.31	4.60	4.50	54.08
1.38	5.05	4.62	57.12
1.44	5.52	4.75	60.25
1.50	6.01	4.88	63.46
1.56	6.52	4.94	65.10
1.62	7.05	5.00	66.76
1.69	7.60	5.12	70.14
1.75	8.18	5.25	73.60
1.81	8.77	5.38	77.15
1.88	9.39	5.44	78.95
1.94	10.02	5.50	80.78
2.00	10.68	5.62	84.49
2.06	11.36	5.75	88.29
2.12	12.06	5.88	92.17
2.18	12.78	5.94	94.14
2.25	13.52	6.00	96.13
2.31	14.28	6.25	104.3
2.38	15.06	6.50	112.8
2.44	15.87	6.75	121.7
2.50	16.69	7.00	130.9
2.56	17.53	7.50	150.2
2.62	18.40	8.00	170.9
2.69	19.29	9.00	216.3
		10.00	267.0
(Continued)		12.00	384.5

COLD DRAWN SQUARES

LOW CARBON—CASE HARDENING

SIZE IN INCHES	WEIGHT PER FT. IN LBS.
.12	.053
.18	.120
.25	.213
.31	.332
.38	.478
.44	.651
.50	.850
.56	1.08
.62	1.33
.69	1.61
.75	1.91
.88	2.60
.94	2.99
1.00	3.40
1.06	3.84
1.12	4.30
1.25	5.31
1.31	5.86
1.38	6.43
1.50	7.65
1.62	8.98
1.75	10.41
1.88	11.95
2.00	13.60
2.12	15.35
2.25	17.21
2.38	19.18
2.50	21.25
2.62	23.43
2.75	25.71
3.00	30.60
3.25	35.91
3.50	41.65
3.75	47.81
4.00	54.40
4.50	68.85
5.00	85.00
6.00	122.4

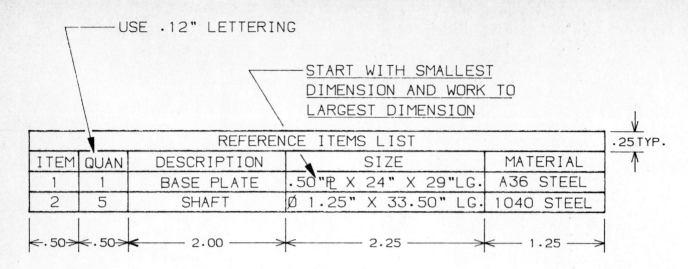

USE .12" LETTERING

START WITH SMALLEST DIMENSION AND WORK TO LARGEST DIMENSION

.25 TYP.

REFERENCE ITEMS LIST				
ITEM	QUAN	DESCRIPTION	SIZE	MATERIAL
1	1	BASE PLATE	.50"℞ X 24" X 29"LG.	A36 STEEL
2	5	SHAFT	Ø 1.25" X 33.50" LG.	1040 STEEL

←.50→ ←.50→ ← 2.00 → ← 2.25 → ← 1.25 →

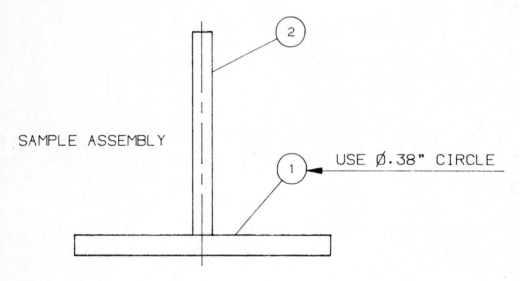

SAMPLE ASSEMBLY

USE Ø.38" CIRCLE

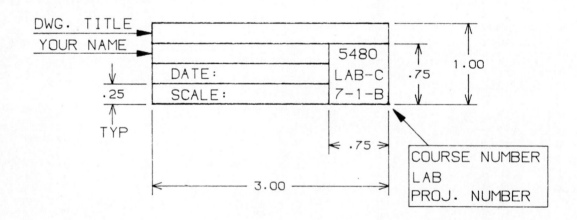

DWG. TITLE

YOUR NAME

	5480
DATE:	LAB-C
SCALE:	7-1-B

.25 TYP

.75 1.00

← .75 →

← 3.00 →

COURSE NUMBER
LAB
PROJ. NUMBER

BOST-BRONZ OIL-IMPREGNATED SINTERED BRONZE BEARINGS

CORED BARS

BOST-BRONZ is stocked in these convenient Bar forms for ease in machining to required bearing size or shape — at your service for all emergencies.

ALL DIMENSIONS IN INCHES
ORDER BY CATALOG NUMBER OR ITEM CODE

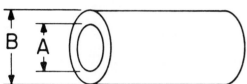

A	B	Catalog No.	Item Code
1/2	1	CB816	35402
	1-1/4	CB820	35404
	1-1/2	CB824	35406
5/8	1	CB1016	35408
	1-1/4	CB1020	35410
	1-3/8	CB1022	35412
	1-1/2	CB1024	35414
	1-3/4	CB1028	35416
3/4	1-1/4	CB1220	35418
	1-1/2	CB1224	35420
	1-3/4	CB1228	35422
	2	CB1232	35424
	2-1/2	CB1240	35426
7/8	1-3/8	CB1422	35428
1	1-1/2	CB1624	35430
	1-3/4	CB1628	35432
	2	CB1632	35434
	2-1/4	CB1636	35436
	2-1/2	CB1640	35438
	3	CB1648	35440
1-1/4	1-3/4	CB2028	35442
	2	CB2032	35444
	2-1/4	CB2036	35446
	2-1/2	CB2040	35448
	3	CB2048	35450
1-3/8	2	CB2232	35452
	2-1/4	CB2236	35454
1-1/2	2	CB2432	35456
	2-1/4	CB2436	35458
	2-1/2	CB2440	35460
	3	CB2448	35462
	3-1/2	CB2456	35464

A	B	Catalog No.	Item Code
1-3/4	2-1/4	CB2836	35466
	2-1/2	CB2840	35468
	2-3/4	CB2844	35470
	3	CB2848	35472
	3-1/2	CB2856	35474
2	2-3/4	CB3244	35476
	3	CB3248	35478
	3-1/4	CB3252	35480
	4	CB3264	35482
	4-1/2	CB3272	35484
	5	CB3280	35486
2-1/4	3	CB3648	35488
	3-1/2	CB3656	35490
	3-3/4	CB3660	35492
2-3/8	3	CB3848	35494
2-1/2	3-1/2	CB4056	35496
3	3-3/4	CB4860	35498
	4	CB4864	35500
	5	CB4880	35502
	6	CB4896	35504
3-1/2	4-3/4	CB5676	35506
3-3/4	5-7/8	CB6094	35508
	8	CB60128	35510
4	6	CB6496	35512
5	7	CB80112	35514

All bars are 6½" long.

STANDARD TOLERANCES

DIMENSIONS		TOLERANCE
A	All	– 1/8"
B	All	+ 1/8"

ANTI-FRICTION BEARINGS

1600 SERIES

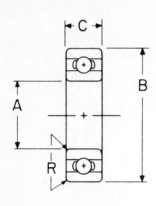

STANDARD TOLERANCES

	DIMENSIONS	TOLERANCE
A	.1875	+ .0000 to – .001
	.2500-1.3125	+ .0000 to – .0005
B	.6875-1.7500	+ .0000 to – .0005
	2.000-2.5625	+ .0000 to – .0006
C	All	+ .000 to – .005

ALL DIMENSIONS IN INCHES

BEARING NUMBER	A	B	C	R Radius*	BALLS No.	BALLS Dia.
1601 1602	.1875 .2500	.6875	1/4 +	.012	6	1/8
1603 1604	.3125 .3750	.8750	9/32 + +	.012 .015	7	5/32
1605 1606 1607	.3125 .3750 .4375	.9062	5/16	.012 .015 .015	9	1/8
1614 1615 1616	.3750 .4375 .5000	1.1250	3/8	.025	7	3/16
1620 1621 1622 1623	.4375 .5000 .5625 .6250	1.3750	7/16	.025	8	15/64
1628 1630	.6250 .7500	1.6250	1/2	.025	8	1/4
1633 1635	.6250 .7500	1.7500	1/2	.025	8	1/4
1638 1640 1641	.7500 .8750 1.0000	2.000	9/16	.035	10	1/4
1652 1654	1.1250 1.2500	2.500	5/8	.035	10	5/16
1657 1658	1.2500 1.3125	2.5625	11/16	.035	9	3/8

* Maximum fillet on shaft or in housing which bearing corner will clear.
+ Width SC & DC = 5/16"
+ + Width SC & DC = 11/32"

LOAD DATA

The indicated load ratings are based on 2500 hours average life. (L$_{50}$) To determine the load ratings at 3500 and 5000 hours. 90 percent and 80 percent respectively, of the above ratings should be used.

BASIC BEARING NUMBER	RADIAL CAPACITY IN POUNDS REVOLUTIONS PER MINUTE								LIMITING THRUST LOADS
	50	100	300	500	1200	1800	2500	5000	
1601 1602	230	185	130	110	80	70	65	50	42
1603 1604	380	300	210	175	130	115	105	80	75
1605 1606 1607	305	245	170	140	105	95	85	65	65
1614 1615 1616	530	420	290	245	185	160	145	115	110
1620 1621 1622 1623	860	690	475	400	300	260	235	185	200
1628 1630 1633 1635	980	780	540	460	340	300	265	210	225
1638 1640 1641	1140	905	630	530	395	345	310	245	280
1652 1654	1695	1345	935	790	590	515	460	365	440
1657 1658	2200	1750	1215	1025	765	665	600	475	570

ANTI-FRICTION BEARINGS

3000 SERIES

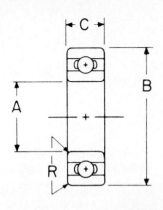

STANDARD TOLERANCES

DIMENSIONS		TOLERANCE
A	All	+ .005 to − .000
B	11/16-1-3/4	+ .0000 to − .0005
	2	+ .0000 to − .0006
C	All	± .005

ALL DIMENSIONS IN INCHES

BEARING NUMBER	A	B	C	R Radius*	BALLS No.	BALLS Dia.
3001	3/16	11/16	1/4 +	.012	6	1/8
3002	1/4					
3003	5/16	7/8	9/32 + +	.012	7	5/32
3004	3/8					
3005	5/16	29/32	5/16	.012	9	1/8
3006	3/8			.015		
3007	7/16			.015		
3014	3/8	1-1/8	3/8	.025	7	3/16
3015	7/16					
3016	1/2					
3020	7/16	1-3/8	7/16	.025	8	15/64
3021	1/2					
3022	9/16					
3023	5/8					
3028	5/8	1-5/8	1/2	.025	8	1/4
3030	3/4					
3033	5/8	1-3/4	1/2	.025	8	1/4
3035	3/4					
3038	3/4	2	9/16	.035	10	1/4
3040	7/8					
3041	1					

*Maximum fillet on shaft or in housing which bearing corner will clear.
+ Width SC & DC = 5/16"
+ + Width SC & DC = 11/32"

LOAD DATA

Load ratings are provided only as a guide for bearing selection and are not to be used for life calculation.

BASIC BEARING NUMBER	RADIAL CAPACITY IN POUNDS REVOLUTIONS PER MINUTE					LIMITING THRUST LOADS
	50	100	500	1800	2500	
3001	150	120	70	45	40	30
3002						
3003	250	200	120	80	70	50
3004						
3005	200	160	95	60	55	43
3006						
3007						
3014	350	280	165	105	95	75
3015						
3016						
3020	575	460	270	175	155	135
3021						
3022						
3023						
3028	650	520	305	200	180	150
3030						
3033						
3035						
3038	760	605	355	230	205	185
3040						
3041						

MOUNTED BALL BEARINGS

PILLOW BLOCKS–LIGHT DUTY
SETSCREW LOCKING

PS SERIES
PRESSED STEEL HOUSING

FEATURES—

Quality pressed steel outer housing.
Deep groove ball bearings for high radial and thrust loads.
Spherical outer race for full self-alignment.
Synthetic lip type seals.
Positive locking by setscrews through extended inner race.
Lubricated for life.

ORDER BY CATALOG
NUMBER OR ITEM CODE

BORE	CATALOG NUMBER	ITEM CODE
1/2	PS-1/2	64500
5/8	PS-5/8	64501
3/4	PS-3/4	64502
7/8	PS-7/8	64503
15/16	PS-15/16	64504
1	PS-1	64505
1- 1/16	PS-1-1/16	64506
1- 1/8	PS-1-1/8	64507
1- 3/16	PS-1-3/16	64508
1- 1/4S	PS-1-1/4S	64509

STANDARD TOLERANCES

DIMENSIONS		TOLERANCE
BORE	All	+ .001 to − .000

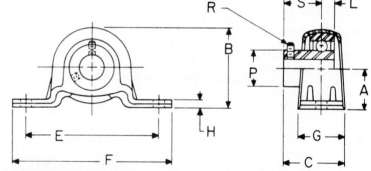

ALL DIMENSIONS IN INCHES

BORE	A	B	C	E	F	G	H	L	P	R* Setscrew (UNF)	S	Bolt Size	Approx. Weight (Lbs.)
1/2 5/8	7/8	1-3/4	1.125	2-11/16	3-5/8	1	.133	15/64	31/32	10 — 32	5/8	5/16	.52 .48
3/4	1	2-1/16	1.203	3	4-1/8	1	.178	9/32	1-11/64	10 — 32	45/64	5/16	.58
7/8 15/16 1	1-1/8	2-7/32	1.328	3- 3/8	4-1/2	1-1/8	.208	19/64	1-11/32	10 — 32	49/64	3/8	.67 .64 .61
1- 1/16 1- 1/8 1- 3/16 1- 1/4S	1-5/16	2-5/8	1.390	3- 3/4	4-7/8	1-1/8	.238	5/16	1-39/64	1/4 — 28	53/64	3/8	1.10 1.05 1.00 .95

* 2 at 120°
Eccentric Locking Collar bearings are available to special order.

MOUNTED BALL BEARINGS

FLANGED UNITS–LIGHT DUTY
SETSCREW LOCKING

PS2/PS3 SERIES
PRESSED STEEL HOUSINGS

ORDER BY CATALOG NUMBER OR ITEM CODE

BORE	3-BOLT		2-BOLT	
	CATALOG NUMBER	ITEM CODE	CATALOG NUMBER	ITEM CODE
1/2	PS3-1/2	64520	PS2-1/2	64510
5/8	PS3-5/8	64521	PS2-5/8	64511
3/4	PS3-3/4	64522	PS2-3/4	64512
7/8	PS3-7/8	64523	PS2-7/8	64513
15/16	PS3-15/16	64524	PS2-15/16	64514
1	PS3-1	64525	PS2-1	64515
1- 1/16	PS3-1-1/16	64526	PS2-1-1/16	64516
1- 1/8	PS3-1-1/8	64527	PS2-1-1/8	64517
1- 3/16	PS3-1-3/16	64528	PS2-1-3/16	64518
1- 1/4S	PS3-1-1/4S	64529	PS2-1-1/4S	64519
1- 1/4	PS3-1-1/4	64530		
1- 5/16	PS3-1-5/16	64531	–	–
1- 3/8	PS3-1-3/8	64532		
1- 7/16	PS3-1-7/16	64533		

STANDARD TOLERANCES

DIMENSIONS		TOLERANCE
BORE	All	+ .001 to – .000

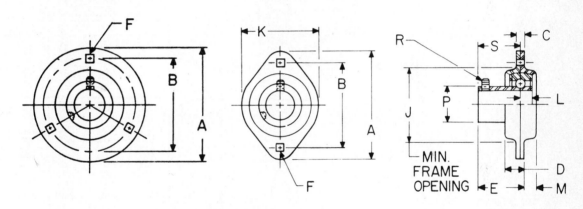

ALL DIMENSIONS IN INCHES

BORE	A	B	C	D	E	F Sq.	J	K	L	M	P	R* Setscrew (UNF)	S	Bolt Size	Approx. Wt. (Lbs.) PS3	PS2
1/2	3- 3/16	2- 1/2	.150	23/64	45/64	9/32	1-15/16	2- 5/16	15/64	13/64	31/32	10 – 32	5/8	1/4	.63	.51
5/8															.59	.47
3/4	3- 9/16	2-13/16	.166	25/64	25/32	11/32	2- 3/16	2- 5/8	9/32	7/32	1-11/64	10 – 32	45/64	5/16	.74	.60
7/8	3- 3/4	3	.166	27/64	27/32	11/32	2- 3/8	2-51/64	19/64	1/4	1-11/32	10 – 32	49/64	5/16	.87	.70
15/16															.84	.67
1															.81	.64
1- 1/16	4- 7/16	3- 9/16	.208	29/64	15/16	13/32	2-13/16	3- 5/16	5/16	1/4	1-39/64	1/4 – 28	53/64	3/8	1.42	1.08
1- 1/8															1.37	1.03
1- 3/16															1.32	.98
1- 1/4S															1.27	.93
1- 1/4	4-13/16	3-15/16	.208	31/64	1- 7/64	13/32	3- 3/16	—	11/32	9/32	1-27/32	1/4 – 28	1	3/8	1.93	
1- 5/16															1.88	
1- 3/8															1.84	–
1- 7/16															1.74	

* 2 at 120°
Eccentric Locking Collar bearings are available to special order. On 1-1/4 through 1-7/16″ Hole Diameters,
Eccentric Collar bearings will have extended inner races, on both sides and will project beyond "M" dimension.

MOUNTED BALL BEARING RADIAL LOAD CAPACITIES
LIGHT DUTY

PS SERIES PILLOW BLOCKS

Shaft Sizes	Average Life (L$_{50}$) Hours	Speed (R.P.M.) Radial Load (Pounds)							
		50	100	500	1000	1500	1800	2000	2500
1/2" 5/8	2,500	300	300	300	300	300	300	300	300
	5,000	300	300	300	300	300	300	300	300
	7,500	300	300	300	245	215	200	195	180
	15,000	300	300	280	220	195	180	175	165
	75,000	300	300	245	195	170	160	155	140
3/4	2,500	350	350	350	350	350	350	350	350
	5,000	350	350	350	350	350	350	350	350
	7,500	350	350	350	330	285	270	260	240
	15,000	350	350	350	300	260	245	235	220
	75,000	350	350	350	260	225	215	205	190
7/8 15/16 1	2,500	400	400	400	400	400	400	400	400
	5,000	400	400	400	400	400	400	400	400
	7,500	400	400	400	360	315	295	285	265
	15,000	400	400	400	325	285	270	260	240
	75,000	400	400	360	285	250	235	225	210
1- 1/16 1- 1/8 1- 3/16 1- 1/4	2,500	600	600	600	600	600	600	600	600
	5,000	600	600	600	600	600	600	600	600
	7,500	600	600	600	500	435	410	395	370
	15,000	600	600	570	455	395	375	360	335
	75,000	600	600	500	395	345	325	315	295

PS2 AND PS3 SERIES FLANGED UNITS

Shaft Sizes	Average Life (L$_{50}$) Hours	Speed (R.P.M.) Radial Load (Pounds)							
		50	100	500	1000	1500	1800	2000	2500
1/2" 5/8	2,500	600	600	600	530	460	435	420	390
	5,000	600	600	530	420	365	345	330	310
	7,500	600	530	310	245	215	200	195	180
	15,000	600	480	280	220	195	180	175	165
	75,000	530	420	245	195	170	160	155	140
3/4	2,500	700	700	700	700	620	585	560	520
	5,000	700	700	700	560	490	460	445	415
	7,500	700	700	415	330	285	270	260	240
	15,000	700	645	375	300	260	245	235	220
	75,000	700	560	330	260	225	215	205	190
7/8 15/16 1	2,500	800	800	800	775	680	640	615	570
	5,000	800	800	775	615	540	505	490	455
	7,500	800	775	455	360	315	295	285	265
	15,000	800	705	410	325	285	270	260	240
	75,000	775	615	360	285	250	235	225	210
1- 1/16 1- 1/8 1- 3/16 1- 1/4S	2,500	1100	1100	1100	1080	940	885	855	795
	5,000	1100	1100	1080	855	750	700	680	630
	7,500	1100	1080	630	500	435	410	395	370
	15,000	1100	980	570	455	395	375	360	335
	75,000	1080	855	500	395	345	325	315	290
1- 1/4 1- 5/16 1- 3/8 1- 7/16	2,500	1400	1400	1400	1400	1245	1175	1130	1050
	5,000	1400	1400	1400	1130	990	930	895	835
	7,500	1400	1400	835	660	580	545	525	485
	15,000	1400	1295	755	600	525	495	475	440
	75,000	1400	1130	660	525	460	430	415	385

Inside Dia		Outside Dia		Width		Inside Dia		Outside Dia		Width	
in.	mm	in.	mm	in.	mm	in.	mm	in.	mm	in.	mm
.375	10	.753	19	.25	6	1.062	26	1.503	38	.38	10
.375	10	.840	21	.31	8		26	1.628	42	.44	12
.438	11	1.003	26	.31	8		26	1.756	44	.44	12
.438	11	1.128	28	.31	8	1.125	28	1.628	42	.44	12
.500	12	1.003	26	.31	8		28	1.756	44	.44	12
.500	12	1.128	28	.31	8		28	1.987	50	.50	12
.500	12	1.254	32	.38	10	1.188	30	1.832	46	.44	12
.562	14	1.003	26	.31	8		30	1.987	50	.50	12
.562	14	1.128	28	.31	8		30	2.254	58	.50	12
.625	16	1.250	32	.38	10	1.25	32	1.756	44	.44	12
.625	16	1.128	28	.31	8		32	1.878	48	.44	12
.625	16	1.250	32	.38	10		32	2.066	52	.50	12
.688	18	1.379	35	.38	10	1.312	34	2.060	52	.44	12
.688	18	1.128	28	.31	8		34	2.254	58	.50	12
.688	18	1.254	32	.38	10		34	2.378	60	.50	12
.750	20	1.379	35	.38	10	1.375	35	2.066	52	.44	12
.750	20	1.254	32	.38			35	2.254	58	.50	12
.750	20	1.379	35		10		35	2.441	62	.50	12
.750	20	1.503	38	.38	10	1.438	36	2.254	58	.50	12
.8125	21	1.756	44	.44	12		36	2.506	64	.50	12
.8125	21	1.254	32	.38	10		36	2.627	66	.50	12
.8125	21	1.379	35	.38	10	1.500	38	2.254	58	.38	10
.8125	21	1.503	38	.38	10		38	2.410	62	.50	12
.875	22	1.756	44	.44	12		38	2.720	70	.50	12
.875	22	1.379	35	.38	10	1.562	40	2.441	62	.50	12
.875	22	1.503	38	.38	10		40	2.690	68	.50	12
.875	22	1.628	42	.44	12		40	2.879	74	.50	12
.938	24	1.756	44	.44	12	1.625	42	2.441	62	.38	10
.938	24	1.503	38	.38	10		42	2.879	74	.38	10
.938	24	1.628	42	.38	10		42	2.627	66	.50	12
1.000	25	1.756	44	.44	12		42	2.879	74	.50	12
1.000	25	1.503	38	.38	10		42	3.066	78	.50	12
1.000	25	1.756	44	.44	12	1.75	44	2.254	58	.50	12
1.000	25	1.878	48	.44	12		44	2.441	62	.50	12
1.000	25	2.004	50	.44	12		44	2.506	64	.50	12

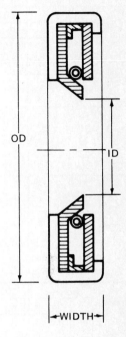

Oil seals.

SPUR GEARS

APPROXIMATE HORSEPOWER AND TORQUE* RATINGS FOR CLASS I SERVICE (Service Factor = 1.0)

12 DIAMETRAL PITCH STEEL   **20° PRESSURE ANGLE**   **1" FACE**   REFERENCE PAGE A41.

No. Teeth	25 RPM		50 RPM		100 RPM		200 RPM		300 RPM		600 RPM		900 RPM		1200 RPM		1800 RPM		3600 RPM	
	H.P.	Torque	H.P.	Torque	H.P.	Torque	H.P.	Torque	H.P.	Torque	H.P.	Torque	H.P.	Torque	H.P.	Torque	H.P.	Torque	H.P.	Torque
12	.08	202	.16	200	.31	196	.60	188	.86	181	1.54	162	2.09	147	2.55	134	3.27	114	4.53	79.4
13	.09	233	.18	230	.36	225	.68	215	.98	206	1.75	183	2.36	165	2.86	150	3.63	127	4.97	87.1
14	.11	265	.21	262	.41	256	.77	244	1.11	233	1.96	206	2.63	184	3.18	167	4.00	140	5.42	94.8
15	.12	297	.23	293	.45	285	.86	271	1.23	259	2.16	227	2.88	202	3.46	182	4.34	152	5.80	102
16	.13	323	.25	319	.49	310	.93	294	1.33	279	2.31	243	3.07	215	3.68	193	4.58	160	6.05	106
18	.15	379	.30	373	.57	361	1.08	340	1.53	322	2.63	276	3.46	242	4.10	215	5.04	177	6.55	115
20	.17	437	.34	429	.66	415	1.23	389	1.74	365	2.95	310	3.84	269	4.52	238	5.50	193	7.02	123
21	.19	468	.36	459	.70	443	1.31	413	1.84	388	3.11	327	4.03	282	4.73	249	5.73	201	7.26	127
24	.22	548	.43	537	.82	515	1.51	477	2.11	444	3.50	368	4.48	314	5.21	274	6.22	218	7.72	135
28	.26	667	.52	651	.99	621	1.80	568	2.49	524	4.04	425	5.10	357	5.86	308	6.89	241		
30	.29	720	.56	707	1.07	673	1.94	612	2.68	562	4.29	451	5.37	376	6.15	323	7.19	252		
36	.37	928	.71	899	1.34	847	2.41	759	3.27	688	5.11	537	6.28	440	7.09	373	8.15	285		
42	.44	1112	.85	1073	1.59	1001	2.81	884	3.77	792	5.73	602	6.94	486	7.76	407				
48	.52	1304	.99	1252	1.84	1159	3.20	1009	4.25	893	6.33	665	7.56	529	8.37	440				
54	.60	1505	1.14	1437	2.09	1319	3.60	1133	4.73	993	6.90	724	8.14	570						

12 DIAMETRAL PITCH CAST IRON   **20° PRESSURE ANGLE**   **1" FACE**   REFERENCE PAGE A42.

No. Teeth	25 RPM		50 RPM		100 RPM		200 RPM		300 RPM		600 RPM		900 RPM		1200 RPM		1800 RPM	3600 RPM
	H.P.	Torque	H.P.	Torque	H.P.	Torque	H.P.	Torque	H.P.	Torque	H.P.	Torque	H.P.	Torque	H.P.	Torque		
60	.40	998	.75	949	1.37	864	2.32	733	3.03	636	4.34	456	5.07	355				
66	.45	1125	.84	1064	1.53	961	2.56	805	3.30	693	4.65	488	5.38	377				
72	.48	1221	.91	1150	1.64	1031	2.71	853	3.47	728	4.81	506	5.53	387				
84	.58	1450	1.07	1354	1.90	1196	3.07	969	3.88	814	5.24	551						
96	.65	1641	1.21	1519	2.10	1322	3.33	1050	4.15	871	5.49	576						
108	.75	1879	1.37	1725	2.35	1482	3.67	1156	4.51	947	5.85	615						
120	.82	2068	1.49	1882	2.53	1596	3.88	1224	4.73	993	6.03	634						
132	.89	2252	1.61	2034	2.70	1704	4.08	1287	4.92	1033	6.19	650						
144	.97	2433	1.73	2181	2.87	1806	4.26	1344	5.09	1070	6.32	664						
168	1.13	2861	2.00	2526	3.25	2047	4.71	1484	5.54	1164								
192	1.27	3209	2.22	2794	3.52	2219	4.99	1573	5.80	1218								
216	1.41	3545	2.42	3045	3.77	2375	5.23	1649	6.01	1263								

Ratings are based on strength calculation. Basic static strength rating, or for hand operation of above gears is approximately 3 times the 100 RPM rating.

NOTE: Ratings to right of heavy line are not recommended, as pitch line velocity exceeds 1000 feet per minute. They should be used for interpolation purposes only.

* Torque Rating (Lb. Ins.).

SPUR GEARS

APPROXIMATE HORSEPOWER AND TORQUE* RATINGS FOR CLASS I SERVICE (Service Factor = 1.0)

10 DIAMETRAL PITCH STEEL   20° PRESSURE ANGLE   1-1/4″ FACE

No. Teeth	25 RPM		50 RPM		100 RPM		200 RPM		300 RPM		600 RPM		900 RPM		1200 RPM		1800 RPM		3600 RPM	
	H.P.	Torque	H.P.	Torque	H.P.	Torque	H.P.	Torque	H.P.	Torque	H.P.	Torque	H.P.	Torque	H.P.	Torque	H.P.	Torque	H.P.	Torque
12	.14	363	.28	358	.55	349	1.06	333	1.51	318	2.66	280	3.57	250	4.30	226	5.40	189	7.27	127
14	.19	477	.37	469	.72	456	1.37	431	1.95	409	3.37	354	4.46	312	5.31	279	6.58	230	8.63	151
15	.21	533	.42	525	.81	509	1.52	479	2.16	453	3.70	389	4.87	341	5.78	303	7.10	249	9.22	161
16	.23	580	.45	571	.88	552	1.64	518	2.32	488	3.96	416	5.18	363	6.12	321	7.47	262	9.60	168
18	.27	679	.53	667	1.02	642	1.90	599	2.67	561	4.48	471	5.79	406	6.79	356	8.19	287	10.33	181
20	.31	784	.61	768	1.17	737	2.16	682	3.02	635	5.00	526	6.41	449	7.45	391	8.90	311	11.04	193
24	.39	983	.76	958	1.45	913	2.65	834	3.65	767	5.89	619	7.41	519	8.50	447	9.98	349		
25	.41	1032	.80	1005	1.52	956	2.76	870	3.80	799	6.10	641	7.64	535	8.74	459	10.21	358		
28	.47	1195	.92	1161	1.74	1097	3.14	990	4.29	901	6.76	710	8.37	586	9.50	499	10.99	385		
30	.52	1300	1.00	1260	1.88	1187	3.38	1064	4.59	964	7.16	752	8.80	616	9.94	522	11.42	400		
35	.64	1615	1.24	1558	2.31	1454	4.08	1284	5.47	1150	8.33	875	10.08	706	11.27	592				
40	.75	1896	1.44	1820	2.67	1685	4.65	1467	6.18	1299	9.20	966	10.99	770	12.17	639				
45	.87	2190	1.66	2092	3.05	1920	5.23	1649	6.88	1445	10.04	1054	11.85	830						
48	.92	2328	1.76	2218	3.21	2026	5.48	1727	7.16	1504	10.33	1085	12.12	849						
50	.96	2420	1.83	2301	3.32	2095	5.64	1777	7.34	1542	10.52	1105	12.29	861						

10 DIAMETRAL PITCH CAST IRON   20° PRESSURE ANGLE   1-1/4″ FACE

No. Teeth	25 RPM		50 RPM		100 RPM		200 RPM		300 RPM		600 RPM		900 RPM		1200 RPM		1800 RPM		3600 RPM	
	H.P.	Torque	H.P.	Torque	H.P.	Torque	H.P.	Torque	H.P.	Torque	H.P.	Torque	H.P.	Torque	H.P.	Torque	H.P.	Torque	H.P.	Torque
55	.65	1638	1.23	1550	2.22	1400	3.72	1173	4.80	1009	6.77	711	7.84	549						
60	.71	1778	1.33	1675	2.38	1501	3.94	1243	5.05	1061	7.01	737	8.06	564						
70	.84	2114	1.57	1974	2.77	1743	4.48	1413	5.65	1187	7.65	803								
80	.98	2462	1.81	2279	3.15	1984	5.00	1576	6.22	1307	8.23	865								
90	1.09	2742	2.00	2517	3.43	2162	5.35	1686	6.58	1382	8.54	897								
100	1.20	3016	2.18	2746	3.70	2329	5.67	1786	6.89	1448	8.80	924								
120	1.45	3650	2.59	3271	4.30	2709	6.40	2016	7.64	1605	9.48	996								
140	1.66	4177	2.93	3688	4.74	2989	6.88	2167	8.09	1699										
160	1.91	4814	3.32	4191	5.28	3329	7.49	2359	8.69	1826										
200	2.30	5802	3.90	4920	5.99	3773	8.17	2573												

8 DIAMETRAL PITCH STEEL   20° PRESSURE ANGLE   1-1/2″ FACE

No. Teeth	25 RPM		50 RPM		100 RPM		200 RPM		300 RPM		600 RPM		900 RPM		1200 RPM		1800 RPM		3600 RPM	
	H.P.	Torque	H.P.	Torque	H.P.	Torque	H.P.	Torque	H.P.	Torque	H.P.	Torque	H.P.	Torque	H.P.	Torque	H.P.	Torque	H.P.	Torque
12	.27	678	.53	667	1.03	647	1.93	609	2.74	576	4.71	495	6.19	434	7.35	386	9.03	316	11.72	205
14	.35	890	.69	874	1.34	843	2.50	787	3.51	738	5.92	622	7.68	537	9.01	473	10.91	382	13.81	242
15	.39	996	.77	976	1.49	939	2.77	873	3.88	816	6.49	681	8.35	585	9.76	513	11.73	411	14.70	257
16	.43	1084	.84	1061	1.62	1018	2.99	943	4.18	877	6.92	727	8.85	620	10.30	541	12.30	431		
18	.50	1268	.98	1238	1.88	1183	3.44	1086	4.77	1003	7.78	817	9.84	689	11.35	596	13.40	469		
20	.58	1462	1.13	1424	2.15	1354	3.91	1233	5.39	1131	8.64	908	10.82	758	12.38	650	14.47	507		
22	.66	1651	1.27	1604	2.41	1518	4.35	1371	5.95	1250	9.41	989	11.67	817	13.27	698	15.36	538		
24	.73	1831	1.41	1775	2.65	1672	4.75	1498	6.46	1357	10.08	1059	12.39	868	14.00	735	16.08	563		
28	.88	2224	1.70	2145	3.18	2003	5.61	1768	7.54	1583	11.47	1204	13.88	972	15.51	815				
32	1.06	2664	2.03	2557	3.76	2367	6.54	2060	8.68	1824	12.92	1358	15.44	1081	17.10	898				
36	1.22	3082	2.34	2944	4.29	2703	7.37	2321	9.68	2034	14.13	1484	16.68	1168						

Ratings are based on strength calculation. Basic static strength rating, or for hand operation of above gears is approximately 3 times the 100 RPM rating.

NOTE: Ratings to right of heavy line are not recommended, as pitch line velocity exceeds 1000 feet per minute. They should be used for interpolation purposes only.

*Torque Rating (Lb. Ins.).

SPUR GEARS

APPROXIMATE HORSEPOWER AND TORQUE* RATINGS
FOR CLASS I SERVICE (Service Factor = 1.0)

8 DIAMETRAL PITCH CAST IRON   20° PRESSURE ANGLE   1-1/2" FACE

No. Teeth	25 RPM H.P.	Torque	50 RPM H.P.	Torque	100 RPM H.P.	Torque	200 RPM H.P.	Torque	300 RPM H.P.	Torque	600 RPM H.P.	Torque	900 RPM H.P.	Torque	1200 RPM H.P.	Torque	1800 RPM H.P.	Torque	3600 RPM H.P.	Torque
40	.84	2111	1.59	2007	2.90	1928	4.92	1550	6.40	1345	9.18	964	10.72	751						
44	.95	2384	1.79	2256	3.23	2038	5.42	1707	6.99	1469	9.86	1035	11.41	799						
48	1.03	2587	1.93	2437	3.47	2184	5.74	1809	7.35	1543	10.20	1072	11.72	821						
56	1.22	3080	2.28	2876	4.03	2539	6.53	2057	8.23	1729	11.14	1170								
60	1.30	3283	2.42	3052	4.25	2676	6.81	2146	8.53	1792	11.40	1198								
64	1.42	3588	2.64	3322	4.59	2892	7.29	2297	9.07	1905	12.00	1260								
72	1.59	3997	2.91	3669	5.00	3151	7.80	2458	9.59	2014	12.44	1307								
80	1.79	4525	3.27	4119	5.54	3493	8.50	2679	10.34	2173	13.20	1386								
88	1.96	4929	3.53	4451	5.92	3729	8.93	2816	10.76	2262	13.54	1422								
96	2.11	5325	3.79	4772	6.27	3952	9.33	2941	11.15	2341	13.83	1453								
112	2.49	6266	4.39	5533	7.11	4483	10.31	3250	12.13	2549										
120	2.64	6651	4.63	5830	7.42	4677	10.63	3351	12.43	2610										
128	2.79	7028	4.85	6118	7.71	4860	10.93	3444	12.69	2667										

6 DIAMETRAL PITCH STEEL   20° PRESSURE ANGLE   2" FACE

No. Teeth	25 RPM H.P.	Torque	50 RPM H.P.	Torque	100 RPM H.P.	Torque	200 RPM H.P.	Torque	300 RPM H.P.	Torque	600 RPM H.P.	Torque	900 RPM H.P.	Torque	1200 RPM H.P.	Torque	1800 RPM H.P.	Torque	3600 RPM H.P.	Torque
12	.63	1559	1.24	1565	2.38	1502	4.41	1391	6.16	1294	10.20	1072	13.06	915	15.19	798	18.14	635		
14	.83	2097	1.62	2046	3.10	1951	5.67	1786	7.84	1647	12.70	1334	16.02	1122	18.42	967	21.67	759		
15	.93	2345	1.81	2284	3.45	2171	6.27	1977	8.64	1814	13.85	1455	17.35	1215	19.85	1043	23.20	812		
16	1.01	2551	1.97	2480	3.73	2351	6.76	2129	9.26	1945	14.71	1545	18.30	1282	20.85	1095	24.21	848		
18	1.18	2981	2.29	2889	4.32	2722	7.74	2440	10.52	2210	16.41	1724	20.18	1413	22.79	1197	26.18	917		
21	1.46	3671	2.81	3541	5.25	3306	9.26	2919	12.44	2613	18.93	1988	22.91	1605	25.61	1345				
24	1.70	4294	3.27	4122	6.05	3815	10.54	3322	14.00	2941	20.83	2188	24.88	1743	27.56	1448				
27	1.98	4986	3.78	4763	6.94	4372	11.92	3755	15.66	3241	22.85	2400	26.98	1889						
30	2.25	5660	4.27	5381	7.77	4899	13.18	4155	17.17	3607	24.60	2584	28.75	2013						

6 DIAMETRAL PITCH CAST IRON   20° PRESSURE ANGLE   2" FACE

No. Teeth	25 RPM H.P.	Torque	50 RPM H.P.	Torque	100 RPM H.P.	Torque	200 RPM H.P.	Torque	300 RPM H.P.	Torque	600 RPM H.P.	Torque	900 RPM H.P.	Torque	1200 RPM H.P.	Torque	1800 RPM H.P.	Torque	3600 RPM H.P.	Torque
33	1.53	3847	2.89	3641	5.22	3288	8.74	2755	11.28	2370	15.90	1670	18.42	1290						
36	1.71	4316	3.23	4066	5.78	3644	9.58	3018	12.26	2575	17.02	1788	19.56	1370						
42	2.04	5148	3.81	4807	6.73	4244	10.91	3439	13.76	2891	18.62	1955								
48	2.38	6009	4.41	5563	7.68	4843	12.21	3847	15.19	3191	20.09	2111								
54	2.74	6899	5.02	6333	8.63	5440	13.46	4243	16.55	3477	21.48	2256								
60	3.01	7591	5.48	6910	9.30	5860	14.26	4494	17.35	3645	22.14	2326								
66	3.38	8515	6.10	7691	10.22	6443	15.44	4864	18.60	3907	23.39	2457								
72	3.65	9200	6.54	8245	10.83	6827	16.12	5080	19.26	4045	23.90	2511								
84	4.30	10835	7.59	9566	12.30	7752	17.83	5620	20.98	4407										
96	4.82	12152	8.39	10579	13.33	8404	18.90	5955	21.95	4611										
108	5.47	13800	9.40	11583	14.67	9245	20.37	6420	23.41	4917										
120	5.97	15059	10.13	12770	15.54	9793	21.20	6680												

Ratings are based on strength calculation. Basic static strength rating, or for hand operation of above gears is approximately 3 times the 100 RPM rating.

NOTE: Ratings to right of heavy line are not recommended, as pitch line velocity exceeds 1000 feet per minute. They should be used for interpolation purposes only.

*Torque Rating (Lb. Ins.).

SPUR GEARS

APPROXIMATE HORSEPOWER AND TORQUE* RATINGS FOR CLASS I SERVICE (Service Factor = 1.0)

5 DIAMETRAL PITCH STEEL **20° PRESSURE ANGLE** **2-1/2" FACE**

No. Teeth	25 RPM H.P.	Torque	50 RPM H.P.	Torque	100 RPM H.P.	Torque	200 RPM H.P.	Torque	300 RPM H.P.	Torque	600 RPM H.P.	Torque	900 RPM H.P.	Torque	1200 RPM H.P.	Torque	1800 RPM H.P.	Torque	3600 RPM H.P.	Torque
12	1.14	2865	2.22	2794	4.22	2662	7.71	2431	10.65	2237	17.19	1805	21.61	1513	24.80	1302	29.09	1019		
14	1.49	3756	2.89	3647	5.47	3449	9.87	3110	13.48	2832	21.25	2233	26.31	1843	29.87	1569	34.53	1209		
15	1.67	4198	3.23	4069	6.08	3833	10.90	3435	14.81	3112	23.11	2427	28.41	1990	32.09	1686	36.87	1291		
16	1.81	4565	3.50	4416	6.58	4146	11.72	3693	15.85	3329	24.47	2570	29.89	2093	33.61	1765				
18	2.12	5332	4.08	5138	7.60	4789	13.38	4216	17.92	3766	27.15	2852	32.77	2295	36.56	1920				
20	2.44	6141	4.68	5894	8.66	5456	15.07	4750	20.02	4205	29.79	3129	35.58	2492	39.41	2070				

5 DIAMETRAL PITCH CAST IRON **20° PRESSURE ANGLE** **2-1/2" FACE**

No. Teeth	25 RPM H.P.	Torque	50 RPM H.P.	Torque	100 RPM H.P.	Torque	200 RPM H.P.	Torque	300 RPM H.P.	Torque	600 RPM H.P.	Torque	900 RPM H.P.	Torque	1200 RPM H.P.	Torque	1800 RPM H.P.	Torque	3600 RPM H.P.	Torque
24	1.82	4599	3.48	4381	6.35	4002	10.82	3411	14.15	2972	20.41	2144	23.95	1677						
25	1.91	4826	3.64	4588	6.63	4177	11.24	3542	14.64	3075	20.97	2203	24.51	1716						
28	2.21	5571	4.18	5267	7.54	4750	12.60	3970	16.23	3410	22.81	2396	26.37	1847						
30	2.40	6050	4.52	5700	8.10	5108	13.42	4230	17.18	3609	23.86	2506	27.41	1920						
35	2.97	7477	5.54	6982	9.78	6164	15.85	4995	19.98	4199	27.04	2840								
40	3.47	8737	6.42	8087	11.17	7040	17.75	5593	22.08	4639	29.21	3068								
45	3.98	10040	7.31	9216	12.56	7916	19.59	6174	24.09	5060	31.26	3284								
50	4.38	11046	7.98	10056	13.53	8528	20.75	6540	25.25	5304	32.22	3384								
60	5.32	13399	9.53	12008	15.78	9944	23.48	7400	28.05	5892	34.81	3657								
70	6.27	15794	11.06	13945	17.93	11300	26.00	8192	30.58	6425										
80	7.23	18229	12.59	15869	20.00	12605	28.34	8932	32.92	6916										
100	8.71	21969	14.78	18630	22.67	14288	30.92	9745												
110	9.68	24409	16.22	20449	24.50	15439	32.88	10362												
120	10.38	26168	17.19	21669	25.58	16125	33.85	10666												
140	11.70	29508	18.97	23910	27.50	17334	35.49	11182												
160	13.30	33526	21.13	26631	29.94	18870	37.83	11921												
180	14.49	36534	22.61	28495	31.40	19787	38.97	12281												

Ratings are based on strength calculation. Basic static strength rating, or for hand operation of above gears is approximately 3 times the 100 RPM rating.

NOTE: Ratings to right of heavy line are not recommended, as pitch line velocity exceeds 1000 feet per minute. They should be used for interpolation purposes only.

*Torque Rating (Lb. Ins.).

SPUR GEARS

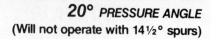

12 *DIAMETRAL PITCH*
STEEL AND CAST IRON

20° *PRESSURE ANGLE*
(Will not operate with 14½° spurs)

ALL DIMENSIONS IN INCHES
ORDER BY CATALOG NUMBER OR ITEM CODE

No. of Teeth	Pitch Dia.	Bore	Hub Dia.	Hub Proj.	Style See Page A93	Without Keyway or Set Screw Catalog Number	Item Code	With Keyway and Setscrew† Catalog Number	Item Code
12 DIAMETRAL PITCH				FACE = 1" OUTSIDE DIA. = PITCH DIA. + .167" OVERALL LENGTH=1"+HUB PROJ.					
STEEL									
12	1.000	1/2	3/4	5/8		YD12	09940	YD12-1/2*	46158
13	1.083		53/64			YD13	09942	YD13-5/8 ‡	46159
14	1.167	5/8	29/32	5/8		YD14	09944	YD14-5/8	46160
15	1.250		63/54			YD15	09946	YD15-5/8	46161
16	1.333		1-1/16			YD16	09948	YD16-5/8	46162
18	1.500	3/4	1-15/64	5/8		YD18	09950	YD18-3/4	46163
20	1.667		1-5/16			YD20	09952	YD20-3/4	46164
21	1.750	3/4 7/8	1-25/64	5/8		YD21	09954	YD21-3/4 YD21-7/8	46165 46166
24	2.000	3/4 7/8 1	1-41/64	5/8	A	YD24 — —	09956 — —	YD24-3/4 YD24-7/8 YD24-1	46167 46168 46169
28	2.333	3/4 7/8 1	1-63/64	5/8		YD28 — —	09958 — —	YD28-3/4 YD28-7/8 YD28-1	46170 46171 46172
30	2.500		2-5/32	5/8		YD30	09960		
36	3.000	3/4	1-15/16			YD36A	10596		
42	3.500		2-7/16	7/8		YD42A	10598		
48	4.000	7/8	2-7/8	7/8		YD48A	10600		
54	4.500		3-3/8			YD54A	10602		

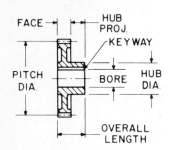

FACE | HUB PROJ.
KEYWAY
PITCH DIA. | BORE | HUB DIA.
OVERALL LENGTH

STANDARD TOLERANCES

DIMENSION		TOLERANCE
BORE	All	± .0005

†3/8" and 1/2" bores have one setscrew.
No keyway.
5/8" bore and larger have standard
keyway at 90° to setscrew.

*YD12-1/2 has 10-32 setscrew.
No keyway.
‡YD13-5/8 has 1/4-20 setscrew.
No keyway.

SPUR GEARS

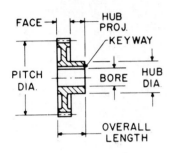

FACE — HUB PROJ. — KEYWAY
PITCH DIA. — BORE — HUB DIA.
OVERALL LENGTH

STANDARD TOLERANCES

DIMENSION		TOLERANCE
BORE	All	± .0005

ALL DIMENSIONS IN INCHES
ORDER BY CATALOG NUMBER OR ITEM CODE

No. of Teeth	Pitch Dia.	Bore	Hub Dia.	Hub Proj.	Style See Page A93	Without Keyway or Set Screw Catalog Number	Without Keyway or Set Screw Item Code	With Keyway and Setscrew† Catalog Number	With Keyway and Setscrew† Item Code
12 DIAMETRAL PITCH						FACE=1" OUTSIDE DIA.=PITCH DIA.+.166" OVERALL LENGTH=1"+HUB PROJ.			
CAST IRON									
60	5.000				B	YD60	10604		
66	5.500					YD66	10606		
72	6.000	7/8	2-1/8	7/8		YD72	10608		
84	7.000					YD84	10610		
96	8.000					YD96	10612		
108	9.000		2-1/4			YD108	10614		
120	10.000		2-1/4	7/8	D	YD120	10616		
132	11.000					YD132	10618		
144	12.000	1	2-1/2	1		YD144	10620		
168	14.000					YD168	10622		
192	16.000					YD192	10624		
216	18.000		2-3/4			YD216	10626		
10 DIAMETRAL PITCH						FACE = 1¼" OUTSIDE DIA. = PITCH DIA. + .200" OVERALL LENGTH=1¼"+HUB PROJ.			
STEEL									
12	1.200	5/8	29/32	5/8		YF12	09963	YF12-5/8	46173
14	1.400		1-7/64			YF14	09964	YF14-5/8	46174
15	1.500	3/4	1-7/32	5/8		YF15	09966	YF15-3/4	46175
16	1.600		1-5/16			YF16	09968	YF16-3/4	46176
18	1.800	3/4 7/8	1-13/32	5/8		YF18 —	09970 —	YF18-3/4 YF18-7/8	46177 46178
20	2.000	7/8 1	1-39/64	5/8		YF20 —	09972 —	YF20-7/8 YF20-1	46179 46180
24	2.400	7/8 1	2-1/64	5/8	A	YF24 —	09974 —	YF24-7/8 YF24-1	46181 46182
25	2.500	7/8 1	2-7/64	5/8		YF25 —	09976 —	YF25-7/8 YF25-1	46183 46184
28	2.800	7/8 1	2-13/32	5/8		YF28 —	09978 —	YF28-7/8 YF28-1	46185 46186
30	3.000	7/8	2	7/8		YF30A	10630		
35	3.500		2-1/2			YF35A	10632		
40	4.000		2-21/64			YF40A	10634		
45	4.500	1	3-29/64	7/8		YF45A	10636		
48	4.800		3-3/4			YF48A	10638		
50	5.000		3-61/64			YF50A	10640		
CAST IRON									
55	5.500				B	YF55	10642		
60	6.000					YF60	10644		
70	7.000	1	2-1/2	1		YF70	10646		
80	8.000					YF80	10648		
90	9.000					YF90	10650		
100	10.000					YF100	10652		
120	12.000	1-1/8	3	1-1/8	D	YF120	10656		
140	14.000					YF140	10658		
160	16.000					YF160	10660		
200	20.000		3-1/4	1-1/4		YF200B	10664		

†All gears have standard keyway at 90° to setscrew.

SPUR GEARS

8 AND **6** DIAMETRAL PITCH
STEEL AND CAST IRON

20° PRESSURE ANGLE
(Will not operate with 14½° spurs)

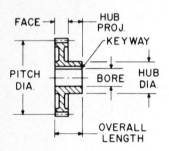

FACE → | ← HUB PROJ.
KEYWAY
PITCH DIA.
BORE — HUB DIA.
OVERALL LENGTH

ALL DIMENSIONS IN INCHES
ORDER BY CATALOG NUMBER OR ITEM CODE

No. of Teeth	Pitch Dia.	Bore	Hub Dia.	Hub Proj.	Style See Page A93	Without Keyway or Set Screw Catalog Number	Without Keyway or Set Screw Item Code	With Keyway and Setscrew† Catalog Number	With Keyway and Setscrew† Item Code
8 DIAMETRAL PITCH						FACE=1½" OUTSIDE DIA.=PITCH DIA.+.250" OVERALL LENGTH=1½"+HUB PROJ.			
STEEL									
12	1.500	3/4	1-1/8	3/4		YH12	09980	YH12-3/4	46187
14	1.750	3/4	1-19/64	3/4		YH14	09982	YH14-3/4	46188
15	1.875	3/4 7/8	1-27/64	3/4		YH15 —	09984 —	YH15-3/4 YH15-7/8	46189 46190
16	2,000	7/8 1	1-35/64	7/8		YH16 —	09986 —	YH16-7/8 YH16-1	46191 46192
18	2.250	7/8 1 1-1/8	1-51/64	7/8		YH18 — —	09988 — —	YH18-7/8 YH18-1 YH18-1-1/8	46193 46194 46195
20	2.500	7/8 1 1-1/8	2-3/64	7/8	A	YH20 — —	09990 — —	YH20-7/8 YH20-1 YH20-1/8	46196 46197 46198
22	2.750	7/8 1 1-1/8	2-19/64	7/8		YH22 — —	09992 — —	YH22-7/8 YH22-1 YH22-1-1/8	46199 46200 46201
24	3.000	7/8 1 1-1/8	2-35/64	7/8		YH24 — —	09994 — —	YH24-7/8 YH24-1 YH24-1-1/8	46202 46203 46204
28	3.500	7/8	3-3/64	7/8		YH28	09996		
32	4.000	1	3	7/8		YH32C	10666		
36	4.500		3-1/2			YH36C	10668		
CAST IRON									
40	5.000				B	YH40B	10670		
44	5.500					YH44B	10672		
48	6.000					YH48B	10674		
56	7.000	1	2-1/2	1	C	YH56B	10676		
60	7.500					YH60	10678		
64	8.000					YH64B	10680		
72	9.000					YH72B	10682		
80	10.000		3		D	YH80B	10684		
88	11.000					YH88B	10686		
96	12.000	1-1/8		1-1/4		YH96B	10688		
112	14.000					YH112B	10690		
120	15.000		3-1/4			YH120	10692		
128	16.000					YH128B	10694		
6 DIAMETRAL PITCH						FACE = 2" OUTSIDE DIA. = PITCH DIA. + .333" OVERALL LENGTH=2"+HUB PROJ.			
STEEL									
12	2.000	1	1-29/64	7/8		YJ12	09998	YJ12-1	46205
14	2.333	1 1-1/8	1-25/32	7/8		YJ14 —	10000 —	YJ14-1 YJ14-1-1/8	46206 46207
15	2.500	1 1-1/8 1-3/16 1-1/4	1-61/64	7/8	A	YJ15 — — —	10002 — — —	YJ15-1 YJ15-1-1/8 YJ15-1-3/16 YJ15-1-1/4	46208 46209 46210 46211

†All gears have standard Keyway at 90° to setscrew.

STANDARD TOLERANCES

DIMENSION		TOLERANCE
BORE	All	± .0005

SPUR GEARS

6 AND **5** *DIAMETRAL PITCH*
STEEL AND CAST IRON

20° *PRESSURE ANGLE*
(Will not operate with 14½° spurs)

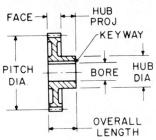

STANDARD TOLERANCES

DIMENSION		TOLERANCE
BORE	All	± .0005

†All gears have standard keyway at 90° to setscrew.

ALL DIMENSIONS IN INCHES
ORDER BY CATALOG NUMBER OR ITEM CODE

No. of Teeth	Pitch Dia.	Bore	Hub Dia.	Hub Proj.	Style See Page A93	Without Keyway or Set Screw Catalog Number	Without Keyway or Set Screw Item Code	With Keyway and Setscrew† Catalog Number	With Keyway and Setscrew† Item Code
6 DIAMETRAL PITCH				FACE=2" OUTSIDE DIA.=PITCH DIA.+.333" OVERALL LENGTH=2"+HUB PROJ.					
STEEL									
16	2.667	1 1-1/8 1-3/16 1-1/4	2-1/8	7/8		YJ16 — — —	10004 — — —	YJ16-1 YJ16-1-1/8 YJ16-1-3/16 YJ16-1-1/4	46212 46213 46214 46215
18	3.000	1 1-1/8 1-3/16 1-1/4	2-29/64	7/8	A	YJ18 — — —	10006 — — —	YJ18-1 YJ18-1-1/8 YJ18-1-3/16 YJ18-1-1/4	46216 46217 46218 46219
21	3.500	1 1-1/8 1-3/16 1-1/4	2-61/64	7/8		YJ21 — — —	10008 — — —	YJ21-1 YJ21-1-1/8 YJ21-1-3/16 YJ21-1-1/4	46220 46221 46222 46223
24 27 30	4.000 4.500 5.000	1-1/8	3 3-1/2 4	7/8		YJ24A YJ27A YJ30C	10704 10706 10708		
CAST IRON									
33 36	5.500 6.000	1-1/8	3 3-1/2	1-1/2	B A	YJ33B YJ36B	10710 10712		
42 48 54	7.000 8.000 9.000	1-1/4	3-1/2	1-1/2	B	YJ42B YJ48B YJ54B	10714 10716 10718		
60 66 72 84 96 108	10.000 11.000 12.000 14.000 16.000 18.000	1-1/4	4	1-1/2	C D	YJ60B YJ66B YJ72B YJ84B YJ96B YJ108B	10720 10722 10724 10726 10728 10730		
120	20.000	1-3/8	4-1/2	1-1/2		YJ120B	10732		
5 DIAMETRAL PITCH				FACE = 2½" OUTSIDE DIA. = PITCH DIA. + .400" OVERALL LENGTH = 2½" + HUB PROJ.					
STEEL									
12 14 15 16 18 20	2.400 2.800 3.000 3.200 3.600 4.000	1-1/8	1-25/32 2-11/64 2-3/8 2-37/64 2-21/32 3-3/8	7/8	A	YK12 YK14 YK15 YK16 YK18 YK20	10010 10012 10014 10016 10018 10020		
CAST-IRON									
24 25 28 30	4.800 5.000 5.600 6.000	1-1/8	3-3/4	1-1/4	A	YK24 YK25B YK28 YK30B	10738 10740 10742 10744		
35 40 45 50	7.000 8.000 9.000 10.000	1-1/4	3-3/4 4	1-1/4	B	YK35B YK40B YK45B YK50B	10746 10748 10750 10752		
60 70 80	12.000 14.000 16.000	1-3/8	4-3/8	1-1/2	D	YK60B YK70B YK80B	10754 10756 10758		
100 110 120	20.000 22.000 24.000	1-1/2	4-3/4 5	1-3/4		YK100B YK110B YK120B	10762 10764 10766		
140 160 180	28.000 32.000 36.000	1-5/8	5 5 5-1/2	2		YK140B YK1608 YK180B	10768 10770 10772		

BEVEL GEARS

STEEL AND IRON — STRAIGHT TOOTH

L Series/PA Series Pinions — Unhardened Steel
PA Series Gears — Cast Iron
HL Series — Hardened Steel (Teeth Only)

APPROXIMATE HORSEPOWER AND TORQUE * RATINGS FOR CLASS I SERVICE (Service Factor = 1.0)

Catalog Number	Pitch	Ratio	50 RPM		100 RPM		200 RPM		300 RPM		600 RPM		900 RPM		1200 RPM		1800 RPM	
			H.P.	Torque	H.P.	Torque	H.P.	Torque	H.P.	Torque	H.P.	Torque	H.P.	Torque	H.P.	Torque	H.P.	Torque
L146Y	16	1-1/2:1	.02	25.2	.03	18.9	.06	18.9	.09	18.9	.16	16.8	.22	15.4	.27	14.18	.35	12.2
HL146Y	16		.03	37.8	.04	25.2	.07	22.0	.11	23.1	.21	22.0	.32	22.4	.42	22.05	.62	21.7
L151Y	12		.06	75.6	.11	69.3	.21	66.2	.30	63.0	.51	53.5	.68	47.6	.80	42.00	1.0	35.0
H151Y	12		.07	88.2	.14	88.2	.28	88.2	.41	86.1	.80	84.0	1.2	84.0	1.5	78.75	2.3	80.5
L153Y	10		.13	164	.25	157	.47	148	.66	139	1.1	115	1.4	98.0	1.6	84.00	1.9	66.5
HL153Y	10		.18	227	.35	220	.68	214	1.0	210	2.0	210	2.9	203	3.8	199.50	5.5	192
L148Y	16	2:1	.01	12.6	.02	12.6	.03	9.4	.05	10.5	.09	9.4	.13	9.1	.15	7.88	.20	7.0
HL148Y	16		.02	25.2	.03	18.9	.04	12.6	.06	12.6	.10	10.5	.14	9.8	.18	9.45	.27	9.4
L149Y	16		.02	25.2	.05	31.5	.09	28.3	.13	27.3	.24	25.2	.32	22.4	.39	20.48	.50	17.5
L150Y	14		.03	37.8	.05	31.5	.10	31.5	.14	29.4	.26	27.3	.35	24.5	.43	22.58	.54	18.9
HL150Y	14		.04	50.4	.06	37.8	.11	34.6	.15	31.5	.29	30.4	.44	30.8	.57	29.93	.84	29.4
HL149Y	16		.03	37.8	.06	37.8	.11	34.6	.16	33.6	.32	33.6	.46	32.2	.61	32.03	.90	31.5
PA3212Y	12		.04	50.4	.09	56.7	.16	50.4	.23	48.3	.40	42.0	.52	36.4	.62	32.55	.76	26.6
PA4212Y	12		.08	101	.16	101	.29	91.3	.41	86.1	.68	71.4	.87	60.9	1.0	52.50	1.2	42.0
L152Y	12		.08	101	.15	94.5	.28	88.2	.40	84.0	.69	72.4	.90	63.0	1.1	57.75	1.3	45.5
PA5210Y	10		.15	189	.28	176	.51	161	.70	147	1.1	115	1.4	98.0	1.6	84.00	—	—
HL152Y	12		.10	126	.19	120	.38	120	.56	118	1.1	115	1.6	112	2.1	110.25	3.1	108
L155Y	10		.16	202	.32	202	.59	186	.82	172	1.4	147	1.7	119	2.0	105.00	2.4	84.0
PA528Y	8		.20	252	.38	239	.70	220	.96	202	1.5	157	1.9	133	2.2	115.50	—	—
L156Y	8		.30	378	.58	365	1.0	315	1.4	294	2.3	241	2.9	203	3.3	173.25	—	—
HL155Y	10		.24	303	.47	296	.93	293	1.4	294	2.7	283	3.9	273	5.2	273.00	7.6	266
PA626Y	8		.40	504	.75	472	1.3	409	1.8	378	2.9	304	3.5	245	4.0	210	—	—
PA625Y	5		.43	542	.82	517	1.5	472	2.0	420	3.1	325	3.8	266	4.3	225.75	—	—
PA726Y	6		.48	605	.89	561	1.6	504	2.1	441	3.9	409	—	—	—	—	—	—
L158Y	6		.59	744	1.1	693	2.0	630	2.7	567	4.2	441	5.2	364	5.9	309.75	—	—
PA826Y	6		.63	794	1.2	756	2.0	630	2.7	567	4.0	420	4.8	336	—	—	—	—
HL156Y	8		.44	555	.88	554	1.7	535	2.5	525	4.9	514	7.3	511	9.5	498.75	13.8	483
PA824Y	4		.98	1235	1.8	1134	3.2	1008	4.2	882	7.3	766	7.5	525	—	—	—	—
HL158Y	6		.77	970	1.5	945	3.0	945	4.4	924	8.5	892	12.5	875	16.4	861.00	23.8	833
PA3116Y	16	3:1	.02	25.2	.04	25.2	.08	25.2	.11	23.1	.20	21.0	.27	18.9	.33	17.33	.42	14.7
PA45312Y	12		.06	75.6	.12	75.6	.22	69.3	.31	65.1	.53	55.6	.70	49.0	.83	43.58	1.0	35.0
PA6310Y	10		.13	164	.24	151	.45	142	.63	132	1.0	105	1.3	91.0	1.5	78.75	1.8	63.0
PA638Y	8		.16	202	.31	195	.57	180	.79	166	1.3	136	1.7	119	2.0	105.00	2.3	80.5
L157Y	10		.19	239	.36	227	.67	211	.94	197	1.6	168	2.0	140	2.3	120.75	2.8	98.0
PA7536Y	6		.33	416	.62	391	1.1	346	1.6	336	2.5	262	3.1	217	3.6	189.00	—	—
PA935Y	5		.57	718	1.1	693	1.9	598	2.6	546	4.1	430	5.0	350	5.7	299.25	—	—
PA4416Y	16	4:1	.02	25.2	.05	31.5	.09	28.3	.13	27.3	.23	24.1	.32	22.4	.38	19.95	.49	17.1
PA6412Y	12		.06	75.6	.12	75.6	.23	72.4	.33	69.3	.56	58.8	.74	51.8	.87	45.68	1.0	35.0
PA6410Y	10		.08	101	.16	101	.30	94.5	.43	90.3	.73	76.6	.92	64.4	1.1	57.75	1.4	49.0
PA848Y	8		.16	202	.31	195	.58	183	.81	170	1.3	136	1.7	119	2.0	105.00	—	—
PA948Y	8		.28	353	.54	340	1.0	315	1.4	294	2.2	231	2.8	196	3.3	173.25	—	—
PA6616Y	16	6:1	.03	37.8	.07	44.1	.13	40.9	.18	37.8	.33	34.6	.44	30.8	.54	28.35	.69	24.1
PA6612Y	12		.05	63.0	.09	56.7	.17	53.5	.25	52.5	.45	47.2	.61	42.7	.75	39.38	.96	33.6

Ratings are based on strength calculation. Basic static strength rating, or for hand operation of above Gears is approximately 3 times the 100 RPM rating.

* Torque Rating (Lb. Ins.)

BEVEL GEARS

STEEL — UNHARDENED AND HARDENED
AND CAST IRON

All gears have "Coniflex"® tooth form.
All Hardened steel gears have teeth only hardened and are equipped with standard keyways and setscrews.

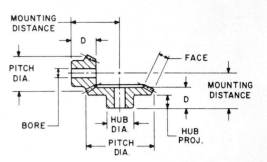

STANDARD TOLERANCES

DIMENSION		TOLERANCE
BORE	All	± .0005

ALL DIMENSIONS IN INCHES
ORDER BY CATALOG NUMBER OR ITEM CODE

Ratio	No. of Teeth	Pitch Dia.	Face	Bore	MD *	D	Hub Dia.	Hub Proj.	Catalog Number (STEEL UNHARDENED)	Item Code	Catalog Number (STEEL HARDENED)	Item Code	Catalog Number (CAST IRON GEARS STEEL PINIONS)	Item Code
16 DIAMETRAL PITCH														
1½:1	24	1.500	.25	1/2	1.188	3/4	1-1/8	9/16	L146Y-G	12230	—		—	
	16	1.000		3/8	1.250	47/64	13/16	7/16	L146Y-P	12232	—		—	
	24	1.500	.25	1/2	1.188	3/4	1-1/8	9/16	—	—	HL146Y-G	11850	—	
	16	1.000		3/8	1.250	47/64	47/64	13/16	—	—	HL146Y-P	11852	—	
2:1	24	1.500	.19	1/2	1.000	5/8	1	7/16	L148Y-G	12238	HL148Y-G	11858	—	
	12	.750		3/8	1.125	37/64	21/32	11/32	L148Y-P	12240	HL148Y-P	11680	—	
	32	2.000	.35	1/2	1.188	49/64	1-1/8	1/2	L149Y-G	12242	HL149Y-G	11862	—	
	16	1.000		3/8	1.500	27/32	13/16	7/16	L149Y-P	12244	HL149Y-P	11864	—	
3:1	48	3.000	.42	5/8	1.312	7/8	1-1/2	9/16	—	—	—	—	PA3316Y-G	12484
	16	1.000		7/16	2.000	59/64	7/8	15/32	—	—	—	—	PA3316Y-P	12486
4:1	64	4.000	.48	5/8	1.375	57/64	2-1/4	9/16	—	—	—	—	PA4416Y-G	12492
	16	1.000		1/2	2.500	63/64	13/16	15/32	—	—	—	—	PA4416Y-P	12494
6:1	96	6.000	.61	5/8	1.688	1-1/4	1-3/16	7/8	—	—	—	—	PA6616Y-G	12516
	16	1.000		1/2	3.750	1-3/8	15/16	23/32	—	—	—	—	PA6616Y-P	12518
14 DIAMETRAL PITCH														
2:1	28	2.000	.35	1/2	1.375	15/16	1-5/8	21/32	L150Y-G	12246	HL150Y-G	11866	—	
	14	1.000	.35	3/8 1/2	1.625	31/32	13/16	9/16	L150Y-G	12248	HL150Y-P	11868	—	
12 DIAMETRAL PITCH														
1½:1	27	2.250	.41	1/2 3/4	1.750	1-1/8	1-1/2	25/32	L151Y-G	12250	HL151Y-G	11870	—	
	18	1.500	.41	1/2	1.875	1-1/8	1-1/4	21/32	L151Y-P	12254	HL151Y-P	11872	—	
2:1	36	3.000	.53	5/8	1.875	1-17/64	2-1/8	7/8	L152BY-G	12260	—	—	—	
	18	1.500		1/2	2.375	1-3/8	1-5/16	13/16	L152BY-P	12262	—	—	—	
	36	3.000	.53	1	1.875	1-17/64	2-1/8	7/8	L152Y-G	12256	HL152Y-G	11874	—	
	18	1.500	.53	5/8 3/4	2.375	1-3/8	1-5/16	13/16	L152Y-P	12258	HL152Y-P	11876	—	

* Mounting Distance (MD) must not be made less than dimension shown.

BEVEL GEARS

12 AND 10 DIAMETRAL PITCH
STEEL — UNHARDENED AND HARDENED
AND CAST IRON

20° PRESSURE ANGLE

All gears have "Coniflex"® tooth form.
All Hardened steel gears have teeth only hardened and are equipped with standard keyways and setscrews.

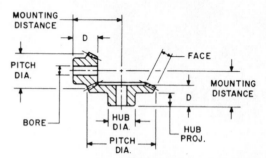

STANDARD TOLERANCES

DIMENSION		TOLERANCE
BORE	All	± .0005

ALL DIMENSIONS IN INCHES
ORDER BY CATALOG NUMBER OR ITEM CODE

Ratio	No. of Teeth	Pitch Dia.	Face	Bore	MD*	D	Hub Dia.	Hub Proj.	Catalog Number	Item Code	Catalog Number	Item Code	Catalog Number	Item Code
12 DIAMETRAL PITCH									STEEL UNHARDENED		STEEL HARDENED		CAST IRON GEARS STEEL PINIONS	
2:1	36	3.000	.46	5/8	1.500	7/8	1-7/16	1/2	—	—	—	—	PA3212Y-G	12480
	18	1.500		1/2	2.250	1-13/64	1-1/4	11/16	—	—	—	—	PA3212Y-P	12482
2:1	48	4.000	.59	5/8	2.000	1-11/64	1-5/8	3/4	—	—	—	—	PA4212Y-G	12488
	24	2.000		1/2	2.875	1-7/16	1-1/2		—	—	—	—	PA4212Y-P	12490
3:1	54	4.500	.60	5/8	1.750	1-1/16	1-3/4	3/4	—	—	—	—	PA45312Y-G	12532
	18	1.500		1/2	3.000	1-11/32	1-1/4	11/16	—	—	—	—	PA45312Y-P	12534
4:1	72	6.000	.60	3/4	2.000	1-5/16	2	61/64	—	—	—	—	PA6412Y-G	12508
	18	1.500		1/2	3.750	1-27/64	1-1/4	23/32	—	—	—	—	PA6412Y-P	12510
6:1	72	6.000	74	3/4	1.750	1-5/16	2	61/64	—	—	—	—	PA6612Y-G	12512
	12	1.000		1/2	3.750	1-31/64	15/16	23/32	—	—	—	—	PA6612Y-P	12514
10 DIAMETRAL PITCH														
1½:1	30	3.000	.57	3/4 1	2.250	1-7/16	2-1/2	1	L153Y-G —	12264 —	— HL153Y-G	— 11878	— —	— —
	20	2.000	.57	3/4	2.500	1-33/64	1-3/4	29/32	L153Y-P	12266	HL153Y-P	11880		
2:1	40	4.000	.71	7/8	2.500	1-11/16	3	1-3/16	L155Y-G	12268	—	—	—	—
	20	2.000		3/4	3.125	1-51/64	1-3/4	1-1/16	L155Y-P	12270	—	—	—	—
2:1	40	4.000	.71	1-1/4	2.500	1-11/16	3	1-3/16	—	—	HL155Y-G	11882	—	—
	20	2.000		7/8	3.125	1-51/64	1-3/4	1-1/16	—	—	HL155Y-P	11884	—	—
	50	5.000	.70	3/4	2.625	1-19/32	2	1	—	—	—	—	PA5210Y-G	12496
	25	2.500			3.375	1-35/64	2	3/4	—	—	—	—	PA5210Y-P	12498
3:1	60	6.000	.78	1	2.750	1-55/64	3	1-3/8	L157Y-G	12274	—	—	—	—
	20	2.000		7/8	4.375	2-5/32	1-3/4	1-5/16	L157Y-P	12276	—	—	—	—
	60	6.000	.78	7/8	2.750	1-29/32	3	1-3/8	—	—	—	—	PA6310Y-G	12500
	20	2.000		3/4	4.375	2-5/32	1-3/4	1-5/16	—	—	—	—	PA6310Y-P	12502
4:1	60	6.000	.72	7/8	2.250	1-5/8	2-1/2	1-1/8	—	—	—	—	PA6410Y-G	12504
	15	1.500		5/8	3.875	1-39/64	1-7/16	27/32	—	—	—	—	PA6410Y-P	12506
6:1	90	9.000	.86	1	2.500	1-13/16	2-3/4	1-5/16	—	—	—	—	PA9610Y-G	12524
	15	1.500		5/8	5.500	1-55/64	1-7/16	31-32	—	—	—	—	PA9610Y-P	12526

* Mounting Distance (MD) must not be made less than dimension shown.

BEVEL GEARS

STEEL – UNHARDENED AND HARDENED AND CAST IRON

All gears have "Coniflex"® tooth form.
All Hardened steel gears have teeth only hardened and are equipped with standard keyways and setscrews.

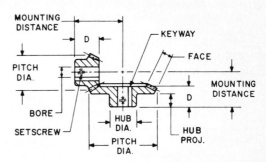

STANDARD TOLERANCES

DIMENSION		TOLERANCE
BORE	All	± .0005

ALL DIMENSIONS IN INCHES
ORDER BY CATALOG NUMBER OR ITEM CODE

Ratio	No. of Teeth	Pitch Dia.	Face	Bore	MD *	D	Hub Dia.	Hub Proj.	Catalog Number STEEL UNHARDENED	Item Code	Catalog Number STEEL HARDENED	Item Code	Catalog Number CAST IRON GEARS STEEL PINIONS	Item Code
8 DIAMETRAL PITCH														
2:1	40	5.000	.82	1 1-1/2	2.875	1-27/32	3	1-1/4	L156Y-G	12252	HL156Y-G	11886	—	—
	20	2.500	.82	1	4.000	2-9/32	2-1/8	1-13/32	L156Y-P	12272	HL156Y-P	11888		
2:1	40	5.000	.82	1	2.875	1-27/32	3	1-1/4	—	—	—	—	PA528Y-G	12424
	20	2.500		7/8	4.000	2-9/32	2-1/8	1-13/32	—	—	—	—	PA528Y-P	12426
3:1	48	6.000	.84	7/8	2.375	1-5/8	2-3/4	1	—	—	—	—	PA638Y-G	12436
	16	2.000		3/4	4.250	2-5/64	1-3/4	1-3/16	—	—	—	—	PA638Y-P	12438
4:1	64	8.000	.84	1	2.750	1-7/8	2-3/4	1-1/4	—	—	—	—	PA848Y-G	12452
	16	2.000		7/8	5.250	2-3/32	1-7/8	1-7/32	—	—	—	—	PA848Y-P	12454
	72	9.000	1.22	1-1/8	3.250	2-5/16	3	1-11/16	—	—	—	—	PA948Y-G	12460
	18	2.250		7/8	5.750	2-15/32	2-1/8	1-7/32	—	—	—	—	PA948Y-P	12462
6 DIAMETRAL PITCH														
2:1	36	6.000	1.06	1-1/8	3.500	2-1/4	3-1/4	1-1/2	L158Y-G	12278	—	—	—	—
	18	3.000			4.750	2-49/64	2-1/2	1-19/32	L158Y-P	12280	—	—	—	—
	36	6.000	1.06	1-3/4	3.500	2-1/4	3-1/4	1-1/2	—	—	HL158Y-G	18890	—	—
	18	3.000		1-1/8	4.750	2-49/64	2-1/2	1-19/32	—	—	HL158Y-P	18892	—	—
2:1	36	6.000	1.06	1-1/8	3.500	2-1/4	3-1/4	1-1/2	—	—	—	—	PA626Y-G	12432
	18	3.000		1	4.750	2-49/64	2-1/2	1-19/32	—	—	—	—	PA626Y-P	12434
	42	7.000	1.05	1-1/8	3.750	2-19/64	3-1/2	1-1/2	—	—	—	—	PA726Y-G	12440
	21	3.500		1	5.000	2-33/64	2-1/2	1-1/4	—	—	—	—	PA726Y-P	12442
	48	8.000	1.17	1-1/8	3.438	1-57/64	3-1/4	1	—	—	—	—	PA826Y-G	12448
	24	4.000		1	5.438	2-35/64	2-5/8	1-1/4	—	—	—	—	PA826Y-P	12450
3:1	45	7.500	1.07	1-1/8	3.000	2-1/8	3-1/4	1-1/4	—	—	—	—	PA7536Y-G	12520
	15	2.500		7/8	5.250	2-9/16	2-1/8	1-7/16	—	—	—	—	PA7536Y-P	12522
5 DIAMETRAL PITCH														
2:1	30	6.000	1.04	1-1/8	3.500	2-1/4	3-1/4	1-3/8	—	—	—	—	PA625Y-G	12428
	15	3.000		1	4.375	2-25/64	2-5/8	1-9/32	—	—	—	—	PA625Y-P	12430
3:1	45	9.000	1.31	1-1/4	3.750	2-1/2	3-3/4	1-11/16	—	—	—	—	PA935Y-G	12456
	15	3.000		1	5.875	2-11/16	2-5/8	1-5/16	—	—	—	—	PA935Y-P	12458
4 DIAMETRAL PITCH														
2:1	32	8.000	1.40	1-1/8	4.250	2-11/16	3-3/4	1-9/16	—	—	—	—	PA824Y-G	12444
	16	4.000	1.40	1-1/8	6.000	3-11/32	3-1/4	1-13/16	—	—	—	—	PA824Y-P	12446

* Mounting Distance (MD) must not be made less than dimension shown.

⅜ PITCH — NO. 35 ASA STANDARD ROLLER CHAIN

Revolutions Per Minute — Small Sprocket

No. of Teeth Small Spkt.	100	500	900	1200	1800	2500	3000	3500	4000	4500	5000	5500	6000	6500	7000	7500	8000
17	.29	1.25	2.12	2.75	3.95	5.31	5.63	4.47	3.66	3.06	2.62	2.27	1.99	1.77	1.58	1.42	1.29
18	.31	1.33	2.25	2.92	4.20	5.65	6.13	4.87	3.98	3.34	2.85	2.47	2.17	1.92	1.72	1.55	1.41
19	.33	1.41	2.39	3.10	4.46	5.99	6.65	5.28	4.42	3.62	3.09	2.68	2.35	2.09	1.87	1.68	1.53
20	.35	1.49	2.53	3.27	4.71	6.33	7.18	5.70	4.67	3.91	3.34	2.90	2.54	2.25	2.02	1.82	1.65
21	.37	1.57	2.66	3.45	4.97	6.68	7.73	6.13	5.02	4.21	3.59	3.11	2.73	2.42	2.17	1.96	1.77
22	.39	1.65	2.80	3.63	5.22	7.02	8.27	6.58	5.38	4.51	3.85	3.34	2.93	2.60	2.33	2.10	1.90
23	.41	1.73	2.94	3.81	5.48	7.37	8.68	7.03	5.75	4.82	4.12	3.57	3.13	2.78	2.49	2.24	2.03
24	.43	1.81	3.08	3.98	5.74	7.71	9.09	7.49	6.13	5.14	4.39	3.80	3.34	2.96	2.65	2.39	2.17
25	.44	1.89	3.21	4.16	6.00	8.06	9.50	7.97	6.52	5.47	4.67	4.05	3.55	3.15	2.82	2.54	2.31
28	.50	2.14	3.63	4.71	6.78	9.11	10.7	9.44	7.73	6.48	5.53	4.80	4.21	3.73	3.34	3.01	2.73
30	.54	2.31	3.91	5.07	7.30	9.81	11.6	10.5	8.57	7.18	6.14	5.32	4.67	4.14	3.70	3.34	3.03
32	.58	2.47	4.20	5.44	7.83	10.5	12.4	11.5	9.44	7.91	6.76	5.86	5.14	4.56	4.08	3.68	3.34
35	.64	2.72	4.62	5.99	8.63	11.6	13.7	13.2	10.8	9.06	7.73	6.70	5.88	5.22	4.67	4.21	3.82
40	.74	3.15	5.34	6.92	9.96	13.4	15.8	16.1	13.2	11.1	9.45	8.19	7.19	6.37	5.70	5.14	4.67
45	.84	3.57	6.06	7.85	11.3	15.2	17.9	19.2	15.8	13.2	11.3	9.77	8.57	7.60	6.80	6.14	0

½ PITCH — NO. 40 ASA STANDARD ROLLER CHAIN

Revolutions Per Minute — Small Sprocket

No. of Teeth Small Spkt.	50	200	400	600	900	1200	1800	2400	3000	3500	4000	4500	5000	5500	6000	6500	7000
17	.37	1.29	2.40	3.45	4.98	6.45	8.96	5.82	4.17	3.31	2.71	2.27	1.94	1.68	1.47	1.31	1.17
18	.39	1.37	2.55	3.68	5.30	6.86	9.76	6.34	4.54	3.60	2.95	2.47	2.11	1.83	1.60	1.42	1.27
19	.42	1.45	2.71	3.90	5.62	7.27	10.5	6.88	4.92	3.91	3.20	2.68	2.29	1.98	1.74	1.54	1.38
20	.44	1.53	2.86	4.12	5.94	7.69	11.1	7.43	5.31	4.22	3.45	2.89	2.47	2.14	1.88	1.67	1.49
21	.46	1.62	3.02	4.34	6.26	8.11	11.7	7.99	5.72	4.54	3.71	3.11	2.66	2.30	2.02	1.79	1.60
22	.49	1.70	3.17	4.57	6.58	8.52	12.3	8.57	6.13	4.87	3.98	3.34	2.85	2.47	2.17	1.92	1.72
23	.51	1.78	3.33	4.79	6.90	8.94	12.9	9.16	6.55	5.20	4.26	3.57	3.05	2.64	2.32	2.06	1.84
24	.54	1.87	3.48	5.02	7.23	9.36	13.5	9.76	6.99	5.54	4.54	3.80	3.25	2.81	2.47	2.19	1.96
25	.56	1.95	3.64	5.24	7.55	9.78	14.1	10.4	7.43	5.89	4.82	4.04	3.45	2.99	2.63	2.33	0
28	.63	2.20	4.11	5.93	8.54	11.1	15.9	12.3	8.80	6.99	5.72	4.79	4.09	3.55	3.11	2.76	0
30	.68	2.38	4.43	6.38	9.20	11.9	17.2	13.6	9.76	7.75	6.34	5.31	4.54	3.93	3.45	0	
32	.73	2.55	4.75	6.85	9.86	12.8	18.4	15.0	10.8	8.54	6.99	5.86	5.00	4.33	3.80	0	
35	.81	2.80	5.24	7.54	10.9	14.1	20.3	17.2	12.3	9.76	7.99	6.70	5.72	4.96	0		
40	.93	3.24	6.05	8.71	12.5	16.3	23.4	21.0	15.0	11.9	9.76	8.18	6.99	0			
45	1.06	3.68	6.87	9.89	14.2	18.5	26.6	25.1	17.9	14.2	11.7	9.76	0				

Courtesy American Chain Association

HP ratings for single strand roller chain.

⅝ PITCH — NO. 50 ASA STANDARD ROLLER CHAIN

Revolutions Per Minute — Small Sprocket

No. of Teeth Small Spkt.	50	100	300	500	900	1200	1500	1800	2100	2400	2700	3000	3300	3500	4000	4500
17	.72	1.34	3.60	5.69	9.70	12.6	14.3	10.7	8.48	6.95	5.83	4.98	4.32	3.96	3.23	2.71
18	.77	1.43	3.83	6.05	10.3	13.4	15.6	11.7	9.24	7.58	6.35	5.42	4.70	4.31	3.52	2.95
19	.81	1.51	4.06	6.42	10.9	14.2	16.9	12.7	10.0	8.22	6.89	5.88	5.10	4.68	3.82	3.20
20	.86	1.60	4.30	6.78	11.6	15.0	18.2	13.7	10.8	8.87	7.44	6.35	5.51	5.05	4.12	3.45
21	.90	1.69	4.53	7.15	12.2	15.8	19.3	14.7	11.6	9.55	8.01	6.83	5.93	5.44	4.44	3.71
22	.95	1.77	4.76	7.52	12.8	16.6	20.3	15.8	12.5	10.2	8.59	7.33	6.36	5.83	4.76	3.98
23	1.00	1.86	5.00	7.89	13.4	17.4	21.3	16.9	13.3	10.9	9.18	7.83	6.79	6.23	5.08	4.26
24	1.04	1.95	5.23	8.26	14.1	18.3	22.3	18.0	14.2	11.7	9.78	8.34	7.24	6.64	5.42	4.54
25	1.09	2.04	5.47	8.63	14.7	19.1	23.3	19.1	15.1	12.4	10.4	8.88	7.70	7.06	5.76	4.83
28	1.20	2.30	6.18	9.76	16.6	21.6	26.3	22.7	17.9	14.7	12.3	10.5	9.13	8.37	6.83	0
30	1.33	2.42	6.66	10.5	17.9	23.2	28.4	25.1	19.9	16.3	13.7	11.7	10.1	9.28	7.57	0
32	1.42	2.66	7.14	11.3	19.2	24.9	30.4	27.7	21.9	18.0	15.0	12.9	11.1	10.2	8.34	0
35	1.57	2.93	7.86	12.4	21.2	27.4	33.5	31.7	25.1	20.5	17.2	14.7	12.8	11.7	9.55	0
40	1.81	3.38	9.08	14.3	24.4	31.1	38.7	38.7	30.6	25.1	21.0	18.0	15.6	14.3	0	
45	2.06	3.84	10.3	16.3	27.8	36.0	43.9	46.2	36.5	29.9	25.1	21.4	18.6	0		

(a)

No. of Teeth Small Spkt.	¾ PITCH					NO. 60 ASA STANDARD ROLLER CHAIN												
	Revolutions Per Minute — Small Sprocket																	
	50	100	200	500	700	900	1200	1400	1600	1800	2000	2200	2400	2600	2800	3000	3500
17	1.24	2.30	4.31	9.81	13.3	16.7	21.7	18.2	14.8	12.5	10.6	9.18	8.06	7.15	6.40	5.75	4.57
18	1.32	2.45	4.58	10.4	14.1	17.8	23.0	19.8	16.1	13.6	11.5	10.0	8.78	7.79	6.97	6.27	4.98
19	1.40	2.60	4.86	11.1	15.0	18.8	24.4	21.5	17.5	14.7	12.5	10.9	9.52	8.45	7.56	6.80	5.40
20	1.48	2.75	5.13	11.7	15.9	19.9	25.8	23.2	18.9	15.9	13.5	11.7	10.3	9.12	8.17	7.34	5.83
21	1.56	2.89	5.41	12.3	16.7	21.0	27.2	24.9	20.3	17.1	14.5	12.6	11.1	9.82	8.79	7.90	6.27
22	1.64	3.04	5.69	13.0	17.6	22.1	28.6	26.7	21.8	18.4	15.6	13.5	11.9	10.5	9.42	8.47	6.73
23	1.72	3.19	5.97	13.6	18.4	23.2	30.0	28.6	23.3	19.6	16.7	14.4	12.7	11.3	10.1	9.06	7.19
24	1.80	3.34	6.25	14.2	19.3	24.3	31.4	30.4	24.8	20.9	17.7	15.4	13.5	12.0	10.7	9.65	7.66
25	1.88	3.49	6.53	14.9	20.2	25.4	32.9	32.4	26.4	22.3	18.9	16.4	14.4	12.8	11.4	10.3	8.15
28	2.12	3.95	7.38	16.8	22.8	28.7	37.1	38.4	31.3	26.4	22.4	19.4	17.0	15.1	13.5	12.2	9.66
30	2.29	4.25	7.95	18.1	24.6	30.9	40.0	42.6	34.7	29.2	24.8	21.5	18.9	16.8	15.0	13.5	0
32	2.45	4.56	8.53	19.4	26.3	33.1	42.9	46.9	38.2	32.2	27.3	23.7	20.8	18.5	16.5	14.9	0

No. of Teeth Small Spkt.	1 PITCH					NO. 80 ASA STANDARD ROLLER CHAIN												
	Revolutions Per Minute — Small Sprocket																	
	25	50	100	200	300	400	500	700	900	1000	1200	1400	1600	1800	2000	2200	2400
17	1.55	2.88	5.38	10.0	14.5	18.7	22.9	31.0	38.9	37.6	28.6	22.7	18.6	15.6	13.3	11.5	10.1
18	1.64	3.07	5.72	10.7	15.4	19.9	24.4	33.0	41.3	41.0	31.2	24.8	20.3	17.0	14.5	12.6	11.0
19	1.74	3.25	6.07	11.3	16.3	21.1	25.8	35.0	43.8	44.5	33.9	26.9	22.0	18.4	15.7	13.6	12.0
20	1.84	3.44	6.42	12.0	17.2	22.3	27.3	37.0	46.4	48.1	36.6	29.0	23.8	19.9	17.0	14.7	12.9
21	1.94	3.62	6.76	12.6	18.2	23.6	28.8	39.0	48.9	51.7	39.4	31.2	25.6	21.4	18.3	15.9	13.9
22	2.04	3.81	7.11	13.3	19.1	24.8	30.3	41.0	51.4	55.5	42.2	33.5	27.4	23.0	19.6	17.0	14.9
23	2.14	4.00	7.46	13.9	20.0	26.0	31.7	43.0	53.9	59.2	45.1	35.8	29.3	24.6	21.0	18.2	15.9
24	2.24	4.19	7.81	14.6	21.0	27.2	33.3	45.0	56.4	62.0	48.1	38.1	31.2	26.2	22.3	19.4	17.0
25	2.34	4.38	8.17	15.2	21.9	28.4	34.8	47.0	59.0	64.9	51.1	40.6	33.2	27.8	23.8	20.6	8.34
28	2.65	4.94	9.23	17.2	24.8	32.1	39.3	53.2	66.6	73.3	60.6	48.1	39.4	33.0	28.2	24.4	0
30	2.85	5.33	9.94	18.5	26.7	34.6	42.3	57.3	71.8	78.9	67.2	53.3	43.6	36.6	31.2	24.5	0
32	3.06	5.71	10.7	19.9	28.6	37.1	45.3	61.4	77.0	84.7	74.0	58.7	48.1	40.3	34.4	0	

HP ratings for single strand roller chain.

(b)

NO. OF TEETH SMALL SPKT.	REVOLUTIONS PER MINUTE—SMALL SPROCKET																
	10	25	50	100	150	200	300	400	500	600	700	900	1000	1200	1400	1600	1800
17	2.19	5.00	9.33	17.4	25.1	32.5	46.8	60.6	74.1	87.3	89.0	61.0	52.1	39.6	31.5	25.8	21.6
18	2.33	5.32	9.92	18.5	26.7	34.6	49.8	64.5	78.8	92.9	97.0	66.5	56.8	43.2	34.3	28.1	23.5
19	2.47	5.64	10.5	19.6	28.3	36.6	52.8	68.4	83.6	98.5	105	72.1	61.6	46.8	37.2	30.4	25.5
20	2.61	5.96	11.1	20.7	29.9	38.7	55.8	72.2	88.3	104	114	77.9	66.5	50.6	40.1	32.9	27.5
21	2.75	6.28	11.7	21.9	31.5	40.8	58.8	76.2	93.1	110	122	83.8	71.6	54.4	43.2	35.4	29.6
22	2.90	6.60	12.3	23.0	33.1	42.9	61.8	80.1	97.9	115	131	89.9	76.7	58.4	46.3	37.9	16.6
23	3.04	6.93	12.9	24.1	34.8	45.0	64.9	84.0	103	121	139	96.1	82.0	62.4	49.5	40.5	0
24	3.18	7.25	13.5	25.3	36.4	47.1	67.9	88.0	108	127	146	102	87.4	66.5	52.8	43.2	0
25	3.32	7.58	14.1	26.4	38.0	49.3	71.0	91.9	112	132	152	109	92.9	70.7	56.1	45.9	0
26	3.47	7.91	14.8	27.5	39.7	51.4	74.0	95.9	117	138	159	115	98.6	75.0	59.5	48.7	0
28	3.76	8.57	16.0	29.8	43.0	55.7	80.2	104	127	150	172	129	110	83.8	66.5	54.4	
30	4.05	9.23	17.2	32.1	46.3	60.0	86.4	112	137	161	185	143	122	92.9	73.8	42.4	
32	4.34	9.90	18.5	34.5	49.6	64.3	92.6	120	147	173	199	158	135	102	81.3	0	

HORSEPOWER RATINGS STANDARD SINGLE STRAND ROLLER CHAIN—NO. 140—1¾" PITCH

NO. OF TEETH SMALL SPKT.	REVOLUTIONS PER MINUTE—SMALL SPROCKET															
	10	25	50	100	150	200	250	300	350	400	500	600	700	900	1000	1200	1400
17	3.39	7.73	14.4	26.9	38.8	50.3	61.4	72.4	83.2	93.8	115	127	101	69.1	59.0	44.9	35.6
18	3.61	8.23	15.4	28.6	41.3	53.5	65.3	77.0	88.5	99.8	122	138	110	75.2	64.2	48.9	38.8
19	3.82	8.72	16.3	30.4	43.7	56.7	60.3	81.6	93.8	106	129	150	119	81.6	69.7	53.0	42.1
20	4.04	9.22	17.2	32.1	46.2	59.9	73.2	86.3	99.1	112	137	161	128	88.1	75.2	57.2	45.4
21	4.26	9.72	18.1	33.8	48.7	63.1	77.2	91.0	104	118	144	170	138	94.8	80.9	61.6	48.9
22	4.48	10.2	19.1	35.6	51.3	66.4	81.2	95.6	110	124	151	178	148	102	86.8	66.0	52.4
23	4.70	10.7	20.0	37.3	53.8	69.7	85.2	100	115	130	159	187	158	109	92.8	70.6	56.0
24	4.92	11.2	20.9	39.1	56.3	72.9	89.2	105	121	136	166	196	169	116	98.9	75.2	59.7
25	5.14	11.7	21.9	40.8	58.8	76.2	93.2	110	126	142	174	205	180	123	105	80.0	63.5
26	5.37	12.2	22.8	42.6	61.4	79.5	97.2	115	132	148	181	214	190	131	112	84.8	0
28	5.81	13.3	24.7	46.2	66.5	86.2	105	124	143	161	197	232	213	146	125	94.8	0
30	6.26	14.3	26.7	49.7	71.6	92.8	113	134	154	173	212	249	236	162	138	105	0
32	6.71	15.3	28.6	53.3	76.8	99.5	122	143	165	186	227	267	260	178	152	116	

NO. OF TEETH SMALL SPKT.	REVOLUTIONS PER MINUTE—SMALL SPROCKET																
	10	25	50	100	150	200	250	300	350	400	500	600	700	850	900	1000	1200
17	4.92	11.2	20.9	39.1	56.3	72.9	89.1	105	121	136	166	141	112	83.7	75.8	65.6	49.9
18	5.23	11.9	22.3	41.6	59.9	77.6	94.8	112	128	145	177	154	122	91.2	83.7	71.5	54.4
19	5.55	12.7	23.6	44.1	63.5	82.2	101	118	136	153	188	167	132	98.9	90.8	77.5	59.0
20	5.86	13.4	25.0	46.6	67.1	86.9	106	125	144	162	198	180	143	107	98.1	83.7	63.7
21	6.18	14.1	26.3	49.1	70.7	91.6	112	132	152	171	209	194	154	115	105	90.1	68.5
22	6.50	14.8	27.7	51.6	74.4	96.3	118	139	159	180	220	208	165	123	113	96.6	0
23	6.82	15.6	29.0	54.2	78.0	101	124	146	167	189	231	222	176	132	121	103	0
24	7.14	16.3	30.4	56.7	81.7	106	129	152	175	197	241	237	188	140	129	110	0
25	7.46	17.0	31.8	59.3	85.4	111	135	159	183	206	252	252	200	149	137	117	0
26	7.78	17.8	33.1	61.8	89.1	115	141	166	191	215	263	267	212	158	145	124	0
28	8.43	19.2	35.9	67.0	96.5	125	153	180	207	233	285	298	237	177	162	139	0
30	9.08	20.7	38.7	72.2	104	135	165	194	223	251	307	331	263	196	180	154	
32	9.74	22.2	41.5	77.4	111	144	176	208	239	269	329	365	289	216	198	169	

HORSEPOWER RATINGS STANDARD SINGLE STRAND ROLLER CHAIN—NO. 200—2½″ PITCH

NO. OF TEETH SMALL SPKT.	REVOLUTIONS PER MINUTE—SMALL SPROCKET																
	10	15	20	30	40	50	70	100	150	200	250	300	350	400	500	550	600
17	9.02	13.0	16.8	24.2	31.4	38.4	52.0	71.6	103	134	163	193	221	249	222	192	169
18	9.59	13.8	17.9	25.8	33.4	40.8	55.3	76.2	110	142	174	205	235	265	242	209	184
19	10.2	14.6	19.0	27.3	35.4	43.3	58.6	80.8	116	151	184	217	249	281	262	227	199
20	10.7	15.5	20.1	28.9	37.4	45.8	61.9	85.4	123	159	195	229	264	297	283	245	0
21	11.3	16.3	21.1	30.5	39.5	48.2	65.3	90.0	130	168	205	242	278	313	305	264	0
22	11.9	17.2	22.2	32.0	41.5	50.7	68.7	94.6	136	177	216	254	292	330	327	283	0
23	12.5	18.0	23.3	33.6	43.5	53.2	72.0	99.3	143	185	226	267	307	346	349	303	0
24	13.1	18.9	24.4	35.2	45.6	55.7	75.4	104	150	194	237	279	321	362	372	323	0
25	13.7	19.7	25.5	36.8	47.6	58.2	78.8	109	156	203	248	292	335	378	396	343	0
26	14.3	20.6	26.6	38.4	49.7	60.7	82.2	113	163	212	259	305	350	395	420	364	0

MAXIMUM BORE AND HUB DIAMETERS

(WITH STANDARD KEYWAYS)

ALL DIMENSIONS ARE IN INCHES

NO. OF TEETH	.375" PITCH		.500"		.625"		.750"		1.000"	
	MAX. BORE	MAXIMUM HUB DIA.	MAX. BORE	MAXIMUM HUB DIA.	MAX. BORE	MAXIMUM HUB DIA.	MAX. BORE	MAXIMUM HUB DIA.	MAX. BORE	MAXIMUM HUB DIA.
11	.594	.859	.781	1.172	.969	1.469	1.250	1.766	1.625	2.375
12	.625	.984	.875	1.328	1.156	1.672	1.281	2.016	1.781	2.703
13	.750	1.109	1.000	1.500	1.281	1.875	1.500	2.250	2.000	3.016
14	.844	1.234	1.156	1.656	1.312	2.078	1.750	2.500	2.281	3.344
15	.875	1.359	1.250	1.812	1.531	2.281	1.781	2.750	2.406	3.672
16	.969	1.469	1.281	1.984	1.688	2.484	1.969	2.984	2.719	3.984
17	1.094	1.594	1.375	2.141	1.781	2.688	2.219	3.219	2.812	4.312
18	1.219	1.719	1.531	2.297	1.875	2.891	2.281	3.469	3.125	4.641
19	1.250	1.844	1.688	2.453	2.062	3.078	2.438	3.703	3.312	4.953
20	1.281	1.953	1.781	2.625	2.250	3.281	2.688	3.953	3.500	5.281
21	1.312	2.078	1.844	2.781	2.281	3.484	2.812	4.188	3.750	5.594
22	1.438	2.203	1.938	2.938	2.438	3.688	2.938	4.438	3.875	5.922
23	1.563	2.312	2.094	3.094	2.625	3.891	3.125	4.672	4.188	6.234
24	1.688	2.438	2.250	3.266	2.812	4.078	3.250	4.906	4.562	6.562
25	1.750	2.562	2.281	3.422	2.844	4.281	3.375	5.156	4.688	6.875

NO. OF TEETH	1.250" PITCH		1.500"		1.750"		2.000"		2.500"	
	MAX. BORE	MAXIMUM HUB DIA.	MAX. BORE	MAXIMUM HUB DIA.	MAX. BORE	MAXIMUM HUB DIA.	MAX. BORE	MAXIMUM HUB DIA.	MAX. BORE	MAXIMUM HUB DIA.
11	1.969	2.969	2.312	3.578	2.812	4.172	3.281	4.781	3.938	5.984
12	2.281	3.375	2.750	4.062	3.250	4.750	3.625	5.422	4.719	6.797
13	2.531	3.781	3.062	4.547	3.562	5.312	4.062	6.078	5.156	7.609
14	2.688	4.188	3.312	5.031	3.875	5.875	4.688	6.719	5.719	8.422
15	3.219	4.594	3.750	5.516	4.438	6.453	4.875	7.375	6.250	9.219
16	3.281	5.000	4.000	6.000	4.688	7.015	5.500	8.016	7.000	10.031
17	3.656	5.406	4.469	6.484	5.062	7.578	5.688	8.656	7.438	10.422
18	3.781	5.797	4.656	6.969	5.625	8.141	6.250	9.312	8.125	11.172
19	4.188	6.203	4.938	7.453	5.688	8.703	6.875	9.953	9.000	12.438
20	4.594	6.609	5.438	7.938	6.250	9.266	7.000	10.594	9.750	13.250
21	4.688	7.000	5.688	8.422	6.812	9.828	7.750	10.234	10.000	14.047
22	4.875	7.406	5.875	8.891	7.250	10.391	8.375	11.875	10.875	14.844
23	5.312	7.812	6.375	9.375	7.438	10.938	9.000	12.516	11.625	15.656
24	5.688	8.203	6.812	9.859	8.000	11.500	9.625	13.156	13.000	16.453
25	5.719	8.609	7.250	10.344	8.562	12.062	10.250	13.797	13.500	17.250

STANDARD SPROCKET SIZES

#35		#40		#50		#60		#80		#100		#120		#140		#160		#200	
PITCH DIA.	NO. TEETH	PITCH	NO. TEETH	PITCH	NO. TEETH	PITCH	NO. TEETH	PITCH	NO. TEETH	PITCH	NO. TEETH	PITCH	NO. TEETH	PITCH DIA.	NO. TEETH	PITCH DIA.	NO. TEETH	PITCH	NO. TEETH
1.685"	14	1.307"	8	1.633"	8	1.960"	8	2.613"	8	3.266"	8	4.386"	9	5.117"	9	5.226"	8	7.310"	9
1.804	15	1.462	9	1.827	9	2.193	9	2.924	9	3.655	9	4.854	10	5.663	10	5.848	9	8.090	10
1.922	16	1.618	10	2.023	10	2.427	10	3.236	10	4.045	10	5.324	11	6.212	11	6.472	10	8.872	11
2.041	17	1.775	11	2.219	11	2.662	11	3.550	11	4.437	11	5.796	12	6.672	12	7.099	11	9.660	12
2.159	18	1.932	12	2.415	12	2.898	12	3.864	12	4.830	12	6.268	13	7.313	13	7.727	12	10.447	13
2.278	19	2.089	13	2.612	13	3.134	13	4.179	13	5.223	13	6.741	14	7.864	14	8.357	13	11.235	14
2.397	20	2.247	14	2.809	14	3.371	14	4.494	14	5.617	14	7.215	15	8.917	15	8.958	14	12.025	15
2.516	21	2.405	15	3.006	15	3.607	15	4.810	15	6.012	15	7.689	16	8.970	16	9.620	15	12.815	16
2.635	22	2.563	16	3.204	16	3.844	16	5.126	16	6.407	16	8.163	17	9.524	17	10.252	16	13.605	17
2.754	23	2.721	17	3.401	17	4.082	17	5.442	17	6.803	17	8.638	18	10.078	18	10.885	17	14.397	18
2.873	24	2.879	18	3.599	18	4.319	18	5.759	18	7.198	18	9.113	19	10.632	19	11.518	18	15.190	19
2.992	25	3.038	19	3.797	19	4.557	19	6.076	19	7.595	19	9.589	20	11.187	20	12.151	19	15.982	20
3.111	26	3.196	20	3.995	20	4.794	20	6.392	20	7.991	20	10.064	21	11.742	21	12.785	20	16.775	21
3.230	27	3.355	21	4.194	21	5.032	21	6.710	21	8.387	21	10.540	22	12.297	22	13.419	21	17.567	22
3.349	28	3.513	22	4.392	22	5.270	22	7.027	22	8.783	22	11.016	23	12.852	23	14.053	22	18.360	23
3.468	29	3.672	23	4.590	23	5.508	23	7.344	23	9.180	23	11.492	24	13.407	24	14.688	23	19.152	24
3.588	30	3.831	24	4.788	24	5.746	24	7.661	24	9.577	24	11.968	25	13.963	25	15.323	24	19.947	25
3.826	32	3.989	25	4.987	25	5.984	25	7.979	25	9.973	25	12.444	26	14.518	26	15.958	25	20.740	26
3.945	33	4.148	26	5.185	26	6.222	26	8.296	26	10.370	26	12.921	27	15.074	27	16.593	26	21.535	27
4.064	34	4.307	27	5.384	27	6.460	27	8.614	27	10.767	27	13.397	28	15.630	28	17.228	27	22.330	28
4.183	35	4.466	28	5.582	28	6.699	28	8.931	28	11.164	28	13.874	29	16.186	29	17.863	28	23.122	29
4.303	36	4.625	29	5.781	29	6.937	29	9.249	29	11.561	29	14.350	30	16.742	30	18.498	29	23.917	30
4.422	37	4.783	30	5.979	30	7.175	30	9.567	30	11.958	30	14.827	31	17.298	31	19.134	30	25.505	32
4.541	38	4.942	31	6.178	31	7.413	31	9.885	31	12.356	31	15.303	32	17.854	32	19.769	31	26.300	33
4.660	39	5.101	32	6.376	32	7.652	32	10.202	32	12.753	32	15.780	33	18.410	33	20.405	32	27.890	35
4.780	40	5.260	33	6.575	33	7.890	33	10.520	33	13.150	33	16.257	34	18.966	34	21.040	33	28.685	36
4.899	41	5.419	34	6.774	34	8.129	34	10.838	34	13.547	34	16.734	35	19.523	35	21.676	34	31.070	39
5.018	42	5.578	35	6.972	35	8.367	35	11.156	35	13.945	35	17.211	36	20.079	36	22.312	35	31.865	40
5.137	43	5.737	36	7.171	36	8.605	36	11.474	36	14.342	36	17.687	37	20.635	37	22.947	36	33.455	42
5.257	44	5.896	37	7.370	37	8.844	37	11.792	37	14.740	37	18.164	38	21.192	38	23.583	37	35.045	44
5.376	45	6.055	38	7.569	38	9.082	38	12.110	38	15.137	38	18.641	39	21.748	39	24.219	38	35.940	45
5.495	46	6.214	39	7.767	39	9.321	39	12.428	39	15.534	39	19.118	40	22.305	40	24.855	39	38.225	48
5.614	47	6.373	40	7.966	40	9.559	40	12.746	40	15.932	40	19.595	41	22.861	41	25.491	40	39.815	50
5.734	48	6.532	41	8.165	41	9.798	41	13.064	41	16.329	41	20.072	42	23.418	42	26.127	41	40.610	51
5.853	49	6.691	42	8.363	42	10.036	42	13.382	42	16.727	42	20.549	43	23.974	43	26.763	42	42.995	54
5.972	50	6.850	43	8.562	43	10.275	43	13.700	43	17.124	43	21.026	44	24.531	44	27.399	43	44.587	56
6.091	51	7.009	44	8.761	44	10.513	44	14.018	44	17.522	44	21.503	45	25.087	45	28.035	44	46.177	58
6.211	52	7.168	45	8.960	45	10.752	45	14.336	45	17.920	45	21.980	46	25.644	46	28.671	45	46.972	59
6.330	53	7.327	46	9.159	46	10.990	46	14.654	46	18.317	46	22.935	48	26.201	47	29.307	46	47.767	60
6.449	54	7.486	47	9.357	47	11.229	47	14.972	47	18.715	47	23.889	50	26.757	48	29.943	47	50.155	63
6.569	55	7.645	48	9.556	48	11.467	48	15.290	48	19.112	48	24.843	52	27.314	49	30.580	48	50.950	64
6.688	56	7.804	49	9.755	49	11.706	49	15.608	49	19.510	49	25.798	54	27.871	50	31.216	49	51.745	65
6.807	57	7.963	50	9.954	50	11.945	50	15.926	50	19.908	50	26.275	55	28.427	51	31.852	50	54.132	68
6.927	58	8.122	51	10.153	51	12.183	51	16.244	51	20.305	51	27.229	57	28.984	52	32.488	51	55.722	70
7.046	59	8.281	52	10.351	52	12.422	52	16.562	52	20.703	52	28.661	60	29.541	53	33.124	52	57.315	72
7.165	60	8.440	53	10.550	53	12.660	53	16.880	53	21.100	53	30.570	64	30.097	54	33.761	53		
7.284	61	8.599	54	10.749	54	12.899	54	17.198	54	21.498	54	31.047	65	30.654	55	34.397	54		
7.404	62	8.758	55	10.948	55	13.137	55	17.517	55	21.896	55	32.002	67	31.211	56	35.033	55		
7.523	63	8.917	56	11.147	56	13.376	56	17.835	56	22.293	56	32.479	68	31.768	57	35.669	56		
7.642	64	9.076	57	11.346	57	13.615	57	18.153	57	22.691	57	33.434	70	32.324	58	36.306	57		
7.762	65	9.236	58	11.544	58	13.854	58	18.471	58	23.089	58	34.388	72	32.881	59	36.942	58		
7.881	66	9.395	59	11.743	59	14.092	59	18.789	59	23.486	59	36.298	76	33.438	60	37.578	59		
8.000	67	9.554	60	11.942	60	14.331	60	19.107	60	23.880	60	38.207	80	33.995	61	38.215	60		
8.120	68	10.190	64	12.738	64	14.808	62	20.380	64	25.475	64	40.116	84	34.551	62	38.851	61		
8.239	69	10.349	65	12.936	65	15.046	63	20.698	65	25.873	65	42.981	90	35.108	63	39.487	62		
8.358	70	10.826	68	13.533	68	15.285	64	21.653	68	26.668	67	45.844	96	35.667	64	40.124	63		
8.597	72	11.145	70	13.931	70	15.524	65	22.289	70	27.066	68	48.709	102	36.222	65	40.760	64		
9.074	76	11.463	72	14.329	72	16.001	67	22.926	72	27.862	70	53.483	112	36.779	66	41.396	65		
9.552	80	12.099	76	15.124	76	16.240	68	24.198	76	28.657	72	57.301	120	37.336	67	42.033	66		
10.029	84	12.736	80	15.920	80	16.717	70	24.835	78	29.453	74			37.892	68	42.669	67		
		13.372	84	16.715	84									38.449	69				
10.745	90					17.194	72	25.471	80	30.248	76			39.006	70	43.306	68		
11.342	95	14.327	90	17.909	90	18.149	76	26.744	84	31.839	80			40.120	72	43.942	69		
11.461	96	15.122	95	18.903	95	19.103	80	28.654	90	33.430	84			42.347	76	44.578	70		
12.177	102	15.281	96	19.102	96	20.058	84	30.245	95	35.817	90			44.575	80	45.851	72		
13.371	112	16.236	102	20.295	102	21.490	90	30.563	96	37.806	95			46.802	84	48.397	76		
14.326	120	17.828	112	22.285	112	22.683	95	32.473	102	38.203	96			53.485	96	50.943	80		
		19.101	120	23.876	120	22.922	96	35.655	112	40.591	102					53.489	84		
						24.354	102	38.202	120	44.569	112					57.307	90		
						26.742	112			47.752	120								
						28.651	120												

Masterline Take-Up Units

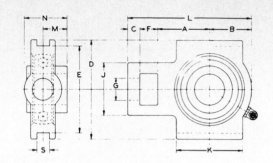

Table No. 12—Masterline Normal Duty Take-up Units—Wide Slot—Setscrew Lock

Shaft Size	Part Number	Use with Frame No.	A	B	C	D	E	F	G	J	K	L	M	N	S	Wt. Lbs.
1/2"	MTWS-208	1SF10	1-3/8"	1-5/16"	3/8"	3-1/2"	3"	5/8"	3/4"	1-7/8"	2"	3-11/16"	5/8"	1-5/16"	1-7/32"	1.8
5/8	MTWS-210	1SF10	1-3/8	1-5/16	3/8	3-1/2	3	5/8	3/4	1-7/8	2	3-11/16	5/8	1-5/16	1-7/32	1.8
3/4	MTWS-212	3 to 9SF16	1-9/16	1-7/16	3/8	3-1/2	3	5/8	3/4	1-7/8	2-1/4	4	23/32	1-13/32	1-7/32	1.8
7/8	MTWS-214	3 to 9SF16	1-17/32	1-9/16	3/8	3-1/2	3	5/8	3/4	1-7/8	2-1/4	4-3/32	1-3/16	1-9/16	1-7/32	2.0
15/16	MTWS-215	3 to 9SF16	1-17/32	1-9/16	3/8	3-1/2	3	5/8	3/4	1-7/8	2-1/4	4-3/32	1-3/16	1-9/16	1-7/32	2.0
1	MTWS-216	3 to 9SF16	1-17/32	1-9/16	3/8	3-1/2	3	5/8	3/4	1-7/8	2-1/4	4-3/32	1-3/16	1-9/16	1-7/32	2.0
1-1/8	MTWS-218	3 to 12SF23	1-19/32	1-25/32	1/2	4	3-1/2	3/4	7/8	2-1/8	2-1/2	4-5/8	7/8	1-3/4	1-7/32	3.1
1-3/16	MTWS-219	3 to 12SF23	1-19/32	1-25/32	1/2	4	3-1/2	3/4	7/8	2-1/8	2-1/2	4-5/8	7/8	1-3/4	1-7/32	3.1
1-1/4	MTWS-220	3 to 12SF23	1-3/4	2	1/2	4	3-1/2	3/4	7/8	2-1/8	2-3/4	5	1	1-7/8	1-7/32	3.7
1-3/8	MTWS-222	3 to 12SF23	1-3/4	2	1/2	4	3-1/2	3/4	7/8	2-1/8	2-3/4	5	1	1-7/8	1-7/32	3.7
1-7/16	MTWS-223	3 to 12SF23	1-3/4	2	1/2	4	3-1/2	3/4	7/8	2-1/8	2-3/4	5	1	1-7/8	1-7/32	3.7
1-1/2	MTWS-224	3 to 18SF31	2-1/32	2-3/16	5/8	4-1/2	4	3/4	1-1/8	2-3/4	3-1/4	5-19/32	1-3/16	2-1/4	1-1/16	5.1
1-5/8	MTWS-226	3 to 18SF31	2-1/8	2-1/4	5/8	4-1/2	4	3/4	1-1/8	2-3/4	3-1/4	5-3/4	1-3/16	2-1/4	1-1/16	5.5
1-11/16	MTWS-227	3 to 18SF31	2-1/8	2-1/4	5/8	4-1/2	4	3/4	1-1/8	2-3/4	3-1/4	5-3/4	1-3/16	2-1/4	1-1/16	5.5
1-3/4	MTWS-228	3 to 18SF31	2-1/8	2-1/4	5/8	4-1/2	4	3/4	1-1/8	2-3/4	3-1/4	5-3/4	1-3/16	2-1/4	1-1/16	5.5
1-15/16	MTWS-231	3 to 18SF31	2-7/32	2-5/16	5/8	4-1/2	4	3/4	1-1/8	2-3/4	3-3/8	5-29/32	1-9/32	2-11/32	1-1/16	5.7
2	MTWS-232	9 to 18SF39	2-17/32	2-9/16	3/4	5-3/4	5-1/8	1	1-3/8	3-5/8	3-3/4	6-27/32	1-5/16	2-9/16	1-1/16	8.6
2-3/16	MTWS-235	9 to 18SF39	2-17/32	2-9/16	3/4	5-3/4	5-1/8	1	1-3/8	3-5/8	3-3/4	6-27/32	1-5/16	2-9/16	1-1/16	8.6
2-7/16	MTWS-239	9 to 18SF39	2-21/32	2-27/32	3/4	5-3/4	5-1/8	1	1-3/8	3-5/8	4	7-1/4	1-9/16	2-13/16	1-1/16	11.0

Part Number Explanation

Masterline Bearing Units

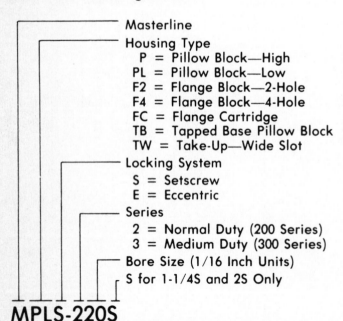

- Masterline
- Housing Type
  - P = Pillow Block—High
  - PL = Pillow Block—Low
  - F2 = Flange Block—2-Hole
  - F4 = Flange Block—4-Hole
  - FC = Flange Cartridge
  - TB = Tapped Base Pillow Block
  - TW = Take-Up—Wide Slot
- Locking System
  - S = Setscrew
  - E = Eccentric
- Series
  - 2 = Normal Duty (200 Series)
  - 3 = Medium Duty (300 Series)
- Bore Size (1/16 Inch Units)
- S for 1-1/4S and 2S Only

MPLS-220S

Masterline Replacement Bearings

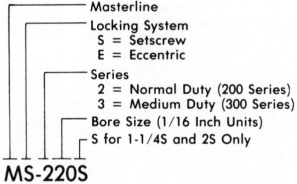

- Masterline
- Locking System
  - S = Setscrew
  - E = Eccentric
- Series
  - 2 = Normal Duty (200 Series)
  - 3 = Medium Duty (300 Series)
- Bore Size (1/16 Inch Units)
- S for 1-1/4S and 2S Only

MS-220S

Take-Up Frames

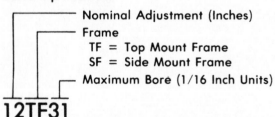

- Nominal Adjustment (Inches)
- Frame
  - TF = Top Mount Frame
  - SF = Side Mount Frame
- Maximum Bore (1/16 Inch Units)

12TF31

Masterline® Take-Up Frames

Normal Duty (200 Series)

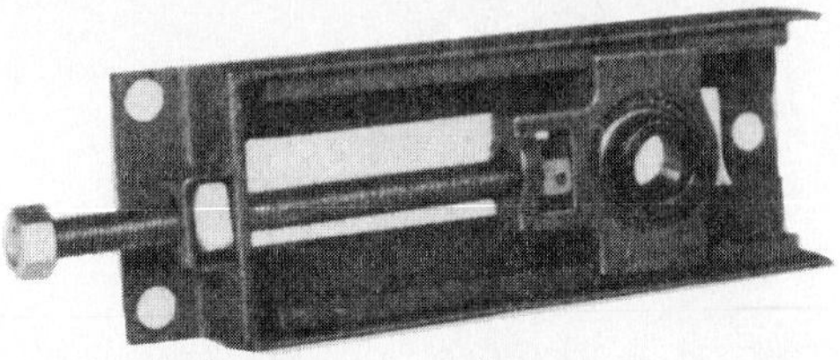

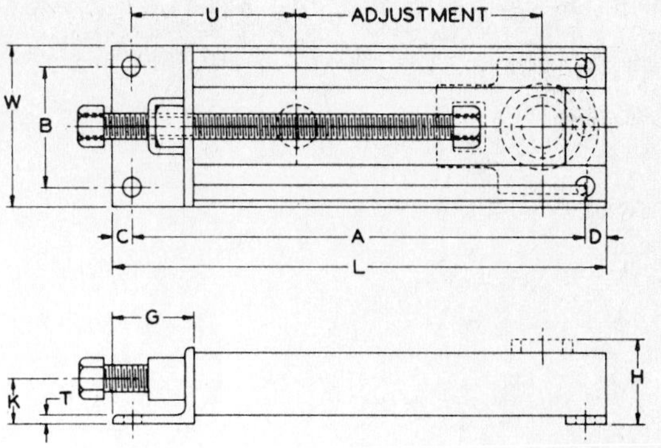

Table No. 13—Side Mount Take-Up Frames

Part Number	Bore Range		Adjust-ment	Dimensions											Bolt Holes		Wt. Lbs.
	200 Series	300 Series		L	W	A	B	C	D	G	H* Max.	K	S▲	T	No.	Size	
1SF10	1/2″-1″	—	1 1/2″	7″	4 1/16″	6″	3 3/16″	1/2″	1/2″	1 1/2″	2 7/16	1 3/8″	5/8″	1/4″	3	9/16	3.3
3SF16	1/2-1	—	3	8 1/2	4 1/16	7 1/2	3 3/16	1/2	1/2	1 1/2	2 7/16	1 3/8	5/8	1/4	3	9/16	3.9
6SF16	1/2-1	—	6	11 1/2	4 1/16	10 1/2	3 3/16	1/2	1/2	1 1/2	2 7/16	1 3/8	5/8	1/4	3	9/16	5.1
9SF16	1/2-1	—	9	14 1/2	4 1/16	13 1/2	3 3/16	1/2	1/2	1 1/2	2 7/16	1 3/8	5/8	1/4	3	9/16	6.2
3SF23	1 3/16-1 7/16	1-1 3/16	3	9 3/8	4 11/16	8 3/8	3 11/16	1/2	1/2	1 1/2	2 21/32	1 3/8	3/4	1/4	3	9/16	5.4
6SF23	1 3/16-1 7/16	1-1 3/16	6	12 3/8	4 11/16	11 3/8	3 11/16	1/2	1/2	1 1/2	2 21/32	1 3/8	3/4	1/4	3	9/16	6.7
9SF23	1 3/16-1 7/16	1-1 3/16	9	15 3/8	4 11/16	14 3/8	3 11/16	1/2	1/2	1 1/2	2 21/32	1 3/8	3/4	1/4	3	9/16	8.1
12SF23	1 3/16-1 7/16	1-1 3/16	12	18 3/8	4 11/16	17 3/8	3 11/16	1/2	1/2	1 1/2	2 21/32	1 3/8	3/4	1/4	3	9/16	9.5
3SF31	1 1/2-1 15/16	1 7/16-1 3/4	3	11	5 1/16	9 7/8	3 1/16	1/2	5/8	1 1/2	3 1/8	1 5/8	1	1/4	4	9/16	7.4
6SF31	1 1/2-1 15/16	1 7/16-1 3/4	6	14	5 1/16	12 7/8	3 1/16	1/2	5/8	1 1/2	3 1/8	1 5/8	1	1/4	4	9/16	9.1
9SF31	1 1/2-1 15/16	1 7/16-1 3/4	9	17	5 1/16	15 7/8	3 1/16	1/2	5/8	1 1/2	3 1/8	1 5/8	1	1/4	4	9/16	10.8
12SF31	1 1/2-1 15/16	1 7/16-1 3/4	12	20	5 1/16	18 7/8	3 1/16	1/2	5/8	1 1/2	3 1/8	1 5/8	1	1/4	4	9/16	12.5
18SF31	1 1/2-1 15/16	1 7/16-1 3/4	18	26	5 1/16	24 7/8	3 1/16	1/2	5/8	1 1/2	3 1/8	1 5/8	1	1/4	4	9/16	15.9
9SF39	2-2 7/16	1 15/16-2 1/4	9	18 1/4	6 3/8	16 7/8	4 1/8	5/8	3/4	2	3 23/32	1 7/8	1 1/4	5/16	4	11/16	20.0
12SF39	2-2 7/16	1 15/16-2 1/4	12	21 1/4	6 3/8	19 7/8	4 1/8	5/8	3/4	2	3 23/32	1 7/8	1 1/4	5/16	4	11/16	22.8
18SF39	2-2 7/16	1 15/16-2 1/4	18	27 1/4	6 3/8	25 7/8	4 1/8	5/8	3/4	2	3 23/32	1 7/8	1 1/4	5/16	4	11/16	28.2

"H" Dimension varies with Take-Up Unit selected. ▲"S" = Take-Up Screw Size.

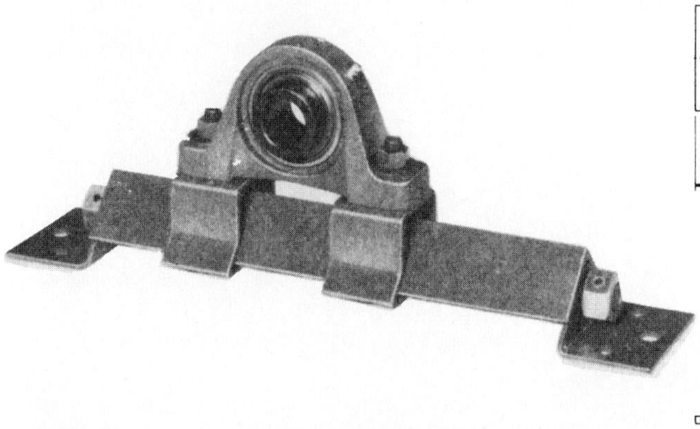

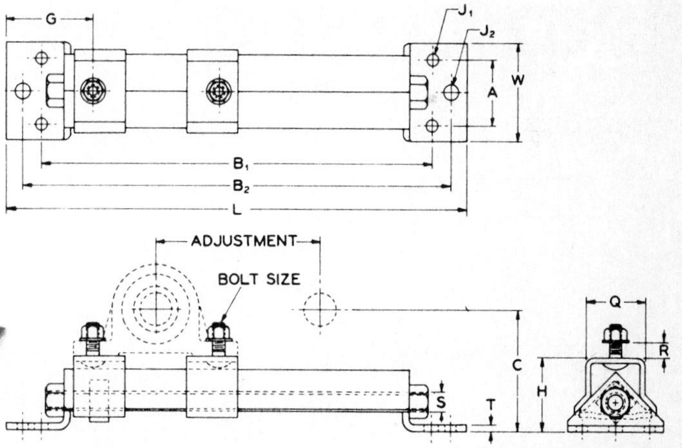

Table No. 14—Top Mount Take-up Frames

Part Number	Bore Range	Nominal Adjust-ment	Dimensions													Bolt Size	Wt. Lbs.
			L	W	A	B1	B2	G	H	J1	J2	Q	R	S	T		
6TF16	3/4″-1″	6″	17 1/8″	3 1/4″	2 1/4″	13 7/8″	15 7/8″	3 5/16″	2 3/8″	7/16″	9/16″	1 3/4″	9/16″	5/8″	3/16″	3/8″	6.0
9TF16	3/4-1	9	20 1/8	3 1/4	2 1/4	16 7/8	18 7/8	3 5/16	2 3/8	7/16	9/16	1 3/4	9/16	5/8	3/16	3/8	6.7
6TF31	1 1/16-1 15/16	6	18 3/16	3 1/2	2 1/2	15 7/16	16 15/16	3 3/8	2 13/16	7/16	9/16	2 5/16	7/8	3/4	3/16	1/2	8.9
9TF31	1 1/16-1 15/16	9	22 5/8	3 1/2	2 1/2	19 7/8	21 3/8	3 3/8	2 13/16	7/16	9/16	2 5/16	7/8	3/4	3/16	1/2	10.3
12TF31	1 1/16-1 15/16	12	24 3/16	3 1/2	2 1/2	21 7/16	22 15/16	3 3/8	2 13/16	7/16	9/16	2 5/16	7/8	3/4	3/16	1/2	10.9
18TF31	1 1/16-1 15/16	18	31 5/8	3 1/2	2 1/2	28 7/8	30 3/8	3 3/8	2 13/16	7/16	9/16	2 5/16	7/8	3/4	3/16	1/2	13.3
12TF39	2-2 7/16	12	28	4 1/2	3	24 3/4	26 3/4	4	3 5/16	9/16	11/16	2 5/8	15/16	1	1/4	5/8	18.3
18TF39	2-2 7/16	18	34	4 1/2	3	30 3/4	32 3/4	4	3 5/16	9/16	11/16	2 5/8	15/16	1	1/4	5/8	21.2

Adjustable Take-Up Application

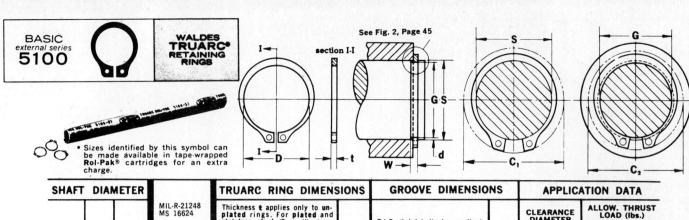

| BASIC external series **5100** | | WALDES **TRUARC** RETAINING RINGS |

• Sizes identified by this symbol can be made available in tape-wrapped **Rol-Pak®** cartridges for an extra charge.

See Fig. 2, Page 45 — section I-I

SHAFT DIAMETER			MIL-R-21248 MS 16624 **EXTERNAL SERIES 5100**	TRUARC RING DIMENSIONS					GROOVE DIMENSIONS					APPLICATION DATA			
						Thickness **t** applies only to un-plated rings. For plated and stainless steel (Type H) rings, add .002″ to the listed maximum thickness. Maximum ring thickness will be at least .0002″ less than the listed minimum groove width (**W**).			T.I.R. (total indicator reading) is the maximum allowable deviation of concentricity between groove and shaft.					CLEARANCE DIAMETER		ALLOW. THRUST LOAD (lbs.) Sharp Corner Abutment	
								Approx. weight per 1000 pieces					Nominal groove depth	When sprung over shaft	When sprung into groove	RINGS (standard material) Safety factor = 4	GROOVES (Cold Rolled steel shafts) Safety factor = 2
				FREE DIA.		THICKNESS			DIAMETER		WIDTH					Important! See Page 25	
Dec. equiv. inch **S**	Approx fract. equiv. inch **S**	Approx mm **S**	size — no.	**D**	tol.	**t**	tol.	**lbs.**	**G**	tol.	**W**	tol.	**d**	**C₁**	**C₂**	**Pr**	**Pg**
.125	⅛	3.2	▲ 5100-12	.112		.010	±.001	.018	.117		.012	+.002 −.000	.004	.222	.214	110	35
.156	5/32	4.0	▲ 5100-15	.142	+.002 −.004	.010		.037	.146	±.0015 .0015 T.I.R.	.012		.005	.270	.260	130	55
.188	3/16	4.8	▲ 5100-18	.168		.015		.059	.175		.018		.006	.298	.286	240	80
.197	— —	5.0	▲ 5100-19	.179		.015		.063	.185		.018		.006	.319	.307	250	85
.219	7/32	5.6	▲ 5100-21	.196		.015		.074	.205		.018		.007	.338	.324	280	110
.236	15/64	6.0	▲ 5100-23	.215		.015		.086	.222		.018		.007	.355	.341	310	120
.250	¼	6.4	• 5100-25	.225		.025		.21	.230		.029		.010	.45	.43	590	175
.276	— —	7.0	5100-27	.250	+.002 −.005	.025		.23	.255		.029		.010	.48	.46	650	195
.281	9/32	7.1	• 5100-28	.256		.025		.24	.261		.029		.010	.49	.47	660	200
.312	5/16	7.9	• 5100-31	.281		.025		.27	.290		.029		.011	.54	.52	740	240
.344	11/32	8.7	5100-34	.309		.025		.31	.321	±.002 .002 T.I.R.	.029		.011	.57	.55	800	265
.354	— —	9.0	5100-35	.320		.025		.35	.330		.029		.012	.59	.57	820	300
.375	⅜	9.5	• 5100-37	.338		.025		.39	.352		.029		.012	.61	.59	870	320
.394	— —	10.0	5100-39	.354		.025		.42	.369		.029		.012	.62	.60	940	335
.406	13/32	10.3	5100-40	.366		.025		.43	.382		.029		.012	.63	.61	950	350
.438	7/16	11.1	• 5100-43	.395		.025		.50	.412		.029		.013	.66	.64	1020	400
.469	15/32	11.9	• 5100-46	.428		.025		.54	.443		.029		.013	.68	.66	1100	450
.500	½	12.7	• 5100-50	.461		.035	±.002	.91	.468	±.002 .004 T.I.R.	.039	+.003 −.000	.016	.77	.74	1650	550
.551	— —	14.0	5100-55	.509		.035		.90	.519		.039		.016	.81	.78	1800	600
.562	9/16	14.3	• 5100-56	.521		.035		1.1	.530		.039		.016	.82	.79	1850	650
.594	19/32	15.1	5100-59	.550		.035		1.2	.559		.039		.017	.86	.83	1950	750
.625	⅝	15.9	• 5100-62	.579		.035		1.3	.588		.039		.018	.90	.87	2060	800
.669	— —	17.0	5100-66	.621		.035		1.4	.629		.039		.020	.93	.89	2200	950
.672	43/64	17.1	5100-66	.621		.035		1.4	.631		.039		.020	.93	.89	2200	950
.688	11/16	17.5	• 5100-68	.635	+.005 −.010	.042		1.8	.646	±.003 .004 T.I.R.	.046		.021	1.01	.97	3400	1000
.750	¾	19.0	• 5100-75	.693		.042		2.1	.704		.046		.023	1.09	1.05	3700	1200
.781	25/32	19.8	5100-78	.722		.042		2.2	.733		.046		.024	1.12	1.08	3900	1300
.812	13/16	20.6	5100-81	.751		.042		2.5	.762		.046		.025	1.15	1.10	4000	1450
.875	⅞	22.2	5100-87	.810		.042		2.8	.821		.046		.027	1.21	1.16	4300	1650
.938	15/16	23.8	5100-93	.867		.042		3.1	.882		.046		.028	1.34	1.29	4650	1850
.984	63/64	25.0	5100-98	.910		.042		3.5	.926		.046		.029	1.39	1.34	4850	2000
1.000	1	25.4	5100-100	.925		.042		3.6	.940		.046		.030	1.41	1.35	4950	2100
1.023	— —	26.0	5100-102	.946		.042		3.9	.961		.046		.031	1.43	1.37	5050	2250
1.062	1 1/16	27.0	5100-106	.982		.050		4.8	.998		.056		.032	1.50	1.44	6200	2400
1.125	1⅛	28.6	5100-112	1.041		.050		5.1	1.059		.056		.033	1.55	1.49	6600	2600
1.188	1 3/16	30.2	5100-118	1.098		.050		5.6	1.118	±.004 .005 T.I.R	.056		.035	1.61	1.54	7000	2950
1.250	1¼	31.7	5100-125	1.156	+.010 −.015	.050		5.9	1.176		.056		.037	1.69	1.62	7350	3250
1.312	1 5/16	33.3	5100-131	1.214		.050		6.8	1.232		.056		.040	1.75	1.67	7750	3700
1.375	1⅜	34.9	5100-137	1.272		.050		7.2	1.291		.056		.042	1.80	1.72	8100	4100
1.438	1 7/16	36.5	5100-143	1.333		.050		8.1	1.350		.056		.044	1.87	1.79	8500	4500
1.500	1½	38.1	5100-150	1.387		.050		9.0	1.406		.056	+.004 −.000	.047	1.99	1.90	9000	5000
1.562	1 9/16	39.7	5100-156	1.446		.062		12.4	1.468		.068		.047	2.10	2.01	11400	5200
1.625	1⅝	41.3	5100-162	1.503		.062		13.2	1.529		.068		.048	2.17	2.08	11850	5500
1.688	1 11/16	42.9	5100-168	1.560		.062		14.8	1.589		.068		.049	2.24	2.15	12350	5850
1.750	1¾	44.4	5100-175	1.618	+.013 −.020	.062	±.003	15.3	1.650	±.005 .005 T.I.R.	.068		.050	2.31	2.21	12800	6200
1.772	— —	45.0	5100-177	1.637		.062		15.4	1.669		.068		.051	2.33	2.23	12950	6400
1.812	1 13/16	46.0	5100-181	1.675		.062		16.2	1.708		.068		.052	2.38	2.28	13250	6650
1.875	1⅞	47.6	5100-187	1.735		.062		17.3	1.769		.068		.053	2.44	2.34	13700	7000
1.969	1 31/32	50.0	5100-196	1.819		.062		18.0	1.857		.068		.056	2.54	2.43	14350	7800
2.000	2	50.8	5100-200	1.850		.062		19.0	1.886		.068		.057	2.55	2.44	14600	8050

▲ Available in beryllium copper only

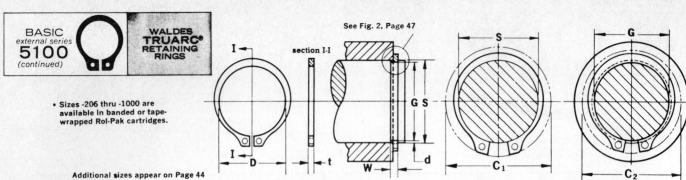

BASIC
external series
5100
(continued)

WALDES
TRUARC®
RETAINING
RINGS

See Fig. 2, Page 47

section I-I

- Sizes -206 thru -1000 are available in banded or tape-wrapped Rol-Pak cartridges.

Additional sizes appear on Page 44

SHAFT DIAMETER			MIL-R-21248 MS 16624 EXTERNAL SERIES **5100**	TRUARC RING DIMENSIONS					GROOVE DIMENSIONS					APPLICATION DATA				
				Thickness **t** applies only to un-plated rings. For plated and stainless steel (Type H) rings, add .002" to the listed maximum thickness. Maximum ring thickness will be at least .0002" less than the listed minimum groove width (**W**).				Approx. weight per 1000 pieces	T.I.R. (total indicator reading) is the maximum allowable deviation of concentricity between groove and shaft.				Nominal groove depth	CLEARANCE DIAMETER		ALLOW. THRUST LOAD (lbs.) Sharp corner abutment	
				FREE DIA.		THICKNESS			DIAMETER		WIDTH			When sprung over shaft	When sprung into groove	RINGS (Standard material) Safety factor = 4	GROOVES (Cold rolled steel shafts) Safety factor = 2
Dec. equiv. inch	Approx. fract. equiv. inch	Approx. mm															Important! See Page 25	
S	S	S	size — no.	D	tol.	t	tol.	lbs.	G	tol.	W	tol.	d	C₁	C₂	Pᵣ	P_g
2.062	2¹⁄₁₆	52.4	5100-206	1.906		.078		25.0	1.946		.086		.058	2.68	2.57	18950	8450
2.125	2⅛	54.0	5100-212	1.964		.078		26.1	2.003		.086		.061	2.75	2.63	19500	9150
2.156	2⁵⁄₃₂	54.8	5100-215	1.993		.078		26.3	2.032		.086		.062	2.78	2.66	19800	9450
2.250	2¼	57.1	5100-225	2.081	+.015 −.025	.078		27.7	2.120		.086		.065	2.87	2.74	20700	10350
2.312	2⁵⁄₁₆	58.7	5100-231	2.139		.078		28.0	2.178		.086		.067	2.94	2.81	21200	10950
2.375	2⅜	60.3	5100-237	2.197		.078		29.2	2.239		.086		.068	3.01	2.88	21800	11400
2.438	2⁷⁄₁₆	61.9	5100-243	2.255		.078		29.5	2.299		.086		.069	3.07	2.94	22400	11900
2.500	2½	63.5	5100-250	2.313		.078		29.7	2.360		.086		.070	3.12	2.98	23000	12350
2.559	− −	65.0	5100-255	2.377		.078		33.9	2.419		.086		.070	3.18	3.04	23500	12650
2.625	2⅝	66.7	5100-262	2.428		.078		35.0	2.481		.086		.072	3.25	3.11	24100	13350
2.688	2¹¹⁄₁₆	68.3	5100-268	2.485		.078		36.0	2.541		.086		.073	3.32	3.18	24700	13850
2.750	2¾	69.8	5100-275	2.543		.093		42.5	2.602		.103		.074	3.45	3.31	30100	14400
2.875	2⅞	73.0	5100-287	2.659		.093		48.5	2.721		.103		.077	3.57	3.42	31500	15650
2.938	2¹⁵⁄₁₆	74.6	5100-293	2.717		.093		50.0	2.779		.103		.079	3.64	3.49	32200	16400
3.000	3	76.2	5100-300	2.775		.093		52.0	2.838		.103		.081	3.69	3.53	32900	17200
3.062	3¹⁄₁₆	77.8	5100-306	2.832		.093		47.5	2.898		.103		.082	3.74	3.58	33500	17750
3.125	3⅛	79.4	5100-312	2.892		.093		58.0	2.957	±.006 .006 T.I.R	.103	+.005 −.000	.084	3.82	3.66	34300	18550
3.156	3⁵⁄₃₂	80.2	5100-315	2.920		.093	±.003	59.0	2.986		.103		.085	3.85	3.68	34600	18950
3.250	3¼	82.5	5100-325	3.006		.093		62.0	3.076		.103		.087	3.95	3.78	35600	20000
3.346	3¹¹⁄₃₂	85.0	5100-334	3.092		.093		64.0	3.166		.103		.090	4.04	3.87	36700	21000
3.438	3⁷⁄₁₆	87.3	5100-343	3.179		.093		66.0	3.257		.103		.090	4.14	3.96	37700	21900
3.500	3½	88.9	5100-350	3.237		.109		72.0	3.316		.120		.092	4.25	4.07	44900	22800
3.543	− −	90.0	5100-354	3.277	+.020 −.030	.109		73.0	3.357		.120		.093	4.29	4.11	45500	23300
3.625	3⅝	92.1	5100-362	3.352		.109		76.0	3.435		.120		.095	4.37	4.18	46600	24300
3.688	3¹¹⁄₁₆	93.7	5100-368	3.410		.109		80.0	3.493		.120		.097	4.43	4.24	47300	25300
3.750	3¾	95.2	5100-375	3.468		.109		83.0	3.552		.120		.099	4.50	4.31	48100	26200
3.875	3⅞	98.4	5100-387	3.584		.109		88.0	3.673		.120		.101	4.60	4.40	49700	27700
3.938	3¹⁵⁄₁₆	100.0	5100-393	3.642		.109		95.0	3.734		.120		.102	4.70	4.50	50600	28400
4.000	4	101.6	5100-400	3.700		.109		101.0	3.792		.120		.104	4.78	4.58	51400	29400
4.250	4¼	108.0	5100-425	3.989		.109		112.0	4.065		.120		.092	5.09	4.91	54600	27600
4.375	4⅜	111.1	5100-437	4.106		.109		115.0	4.190		.120		.092	5.22	5.04	56200	28400
4.500	4½	114.3	5100-450	4.223		.109		101.0	4.310		.120		.095	5.37	5.18	57800	30200
4.750	4¾	120.6	5100-475	4.458		.109		113.0	4.550		.120		.100	5.67	5.47	61000	33600
5.000	5	127.0	5100-500	4.692		.109		149.0	4.790		.120		.105	5.96	5.75	64200	37100
5.250	5¼	133.3	5100-525	4.927		.125		190.0	5.030		.139		.110	6.27	6.05	77300	40800
5.500	5½	139.7	5100-550	5.162	+.020 −.040	.125		202.5	5.265	±.007 .006 T.I.R	.139	+.006 −.000	.117	6.57	6.34	81000	45500
5.750	5¾	146.0	5100-575	5.396		.125	±.004	220.0	5.505		.139		.122	6.86	6.62	84700	49600
6.000	6	152.4	5100-600	5.631		.125		210.0	5.745		.139		.127	7.16	6.91	88300	53800
6.250	6¼	158.7	5100-625	5.866		.156		282.0	5.985		.174		.132	7.46	7.20	114800	58300
6.500	6½	165.1	5100-650	6.100	+.020 −.050	.156		330.0	6.225		.174		.137	7.87	7.60	119400	62900
6.750	6¾	171.4	5100-675	6.335		.156		356.0	6.465		.174		.142	8.06	7.78	124000	67700
7.000	7	177.8	5100-700	6.570		.156		388.0	6.705	±.008 .006 T.I.R	.174	+.008 −.000	.147	8.36	8.07	128600	72700
7.500	7½	190.5	5100-750	7.039		.187	±.005	534.0	7.180		.209		.160	8.96	8.64	165200	84800
8.000	8	203.2	5100-800	7.508		.187		628.0	7.660		.209		.170	9.56	9.22	176200	96100
8.500	8½	215.9	5100-850	7.977	+.020 −.060	.187		700.0	8.140		.209		.180	10.16	9.80	187200	108100
9.000	9	228.6	5100-900	8.445		.187		757.0	8.620		.209		.190	10.75	10.37	198200	120800
9.500	9½	241.3	5100-950	8.915		.187		820.0	9.100		.209		.200	11.34	10.94	209200	134200
10.000	10	254.0	5100-1000	9.385		.187		964.0	9.575		.209		.212	11.94	11.52	220200	149800